# Preface

Intended for class use or self-study, this text aspires to introduce statistical methodology to a wide audience, simply and intuitively, through resampling from the data at hand.

The resampling methods—permutations, cross-validation, and the bootstrap—are easy to learn and easy to apply. They require no mathematics beyond introductory high-school algebra, yet are applicable in an exceptionally broad range of subject areas.

Introduced in the 1930s, the numerous, albeit straightforward, calculations resampling methods require were beyond the capabilities of the primitive calculators then in use. And they were soon displaced by less powerful, less accurate approximations that made use of tables. Today, with a powerful computer on every desktop, resampling methods have resumed their dominant role and table lookup is an anachronism.

Physicians and physicians in training, nurses and nursing students, business persons, business majors, research workers and students in the biological and social sciences will find here a practical and easily-grasped guide to descriptive statistics, estimation, and testing hypotheses.

For advanced students in biology, dentistry, medicine, psychology, sociology, and public health, this text can provide a first course in statistics and quantitative reasoning.

For industrial statisticians, statistical consultants, and research workers, this text provides an introduction and day-to-day guide to the power, simplicity, and versatility of the bootstrap, cross-validation, and permutation tests.

For mathematics majors, this text will form the first course in statistics to be followed by a second course devoted to distribution theory.

Hopefully, all readers will find my objectives are the same as theirs: *To use quantitative methods to characterize, review, report on, test, estimate, and classify findings.*

If you're just starting to use statistics in your work, begin by reading chapters 1 to 5 which cover descriptive statistics, sampling, hypothesis testing, and estimation, along with portions of Chapter 6 with its coverage of the essential, but challenging, ideas of significance level, sample size, and power, and Chapters 7 on contingency tables and/or 8 on experimental design, depending upon your interests. Recurrent themes—for example, the hospital data considered in the exercises for Chapters 1 and 5 to 8—tie the material together and provide a framework for self-study and classes.

For a one-quarter short course, I took the students through Chapter 1 (we looked at, but did not work through, section 1.2 on charts and graphs), Chapter 2, letting the students come up with their own examples and illustrations, Chapter 3 on hypothesis testing, Chapter 4, and Sections 5.1, 5.2, and 5.5 on bootstrap estimation. One group wanted to talk about sample size (Sections 6.2–6.4), the next about contingency tables (Sections 7.1, 7.3, 7.5).

Research workers, familiar with the material in Chapters 1 and 2, should read chapters 3, 5, and 6, and then any and all of Chapters 7 to 9 according to their needs. If you have data in hand, turn first to Chapter 10 whose expert system will guide you to the appropriate sections of the text.

A hundred or more exercises included at the end of each chapter plus dozens of thought-provoking questions will serve the needs of both classroom and self-study. C++ algorithms, Stata, and SC code and a guide to off-the-shelf resampling software are included as appendixes. The reader is invited to download a self-standing IBM-PC program from http://users.oco.net/drphilgood/resamp.htm that will perform most of the permutation tests and simple bootstraps described here. The software is self-guiding, so if you aren't sure what method to use, let the program focus and limit your selection. To spare you and your students the effort of retyping, I've included some of the larger data sets, notably the hospital data (section 1.8) along with the package.

My thanks to Symantek, Design Science Cytel Software, Stata, and Salford Systems without whose GrandView® outliner, Mathtype© equation generator, StatXact® Stata®, and CART® statistics packages this text would not have been possible.

I am deeply indebted to my wife Dorothy, to Bill Sribney for his contributions, to Aric Agmon, Barbara Heller, Tim Hesterberg, David Howell, John Kimmel, Alson Look, Lloyd S. Nelson, Dan Nordlund, Richard Saba, Lazar Tenjovic, and Bill Teel for their comments and corrections, and to the many readers of the first edition who encouraged me to expand and revise this text.

Huntington Beach, California                    *Phillip I. Good*
                                        brother_unknown@yahoo.com

# CHAPTER 1

# Descriptive Statistics

## 1.1 Statistics

Statistics help you

- decide what data and how much data to collect
- analyze your data
- determine the degree to which you may rely on your findings.

A time-worn business adage is to begin with your reports and work back to the data you need. In this chapter, you learn to report your results through graphs, charts, and summary statistics, and to use a sample to describe and estimate the characteristics of the population from which the sample is drawn.

## 1.2 Reporting Your Results

Imagine you are in the sixth grade and you have just completed measuring the heights of all your classmates.

Once the pandemonium has subsided,[1] your instructor asks you and your team to prepare a report summarizing your results.

Actually, you have two sets of results. The first set consists of the measurements you made of you and your team members, reported in centimeters, 148.5, 150.0, and 153.0. (Kelly is the shortest incidentally, while you are the tallest.) The instructor asks you to report the *minimum*, the *median*, and the *maximum* height in your group. This part is easy, or at least it's easy once you look up median in the glossary of your textbook and discover it means "the one in the

---

[1] I spent the fall of 1994 as a mathematics and science instructor at St. John's Episcopal School in Rancho Santa Marguarite, CA. The results reported here, especially the pandemonium, were obtained by my sixth-grade homeroom. We solved the problem of a metric tape measure by building our own from string and a meter stick.

middle." In your group, the minimum height is 148.5 centimeters, the median is 150.0 centimeters, the maximum is 153.0 centimeters. Minimum means smallest, maximum means largest, median is the one in the middle. Conscientiously, you write these definitions down—they could be on a test.[2]

Your second assignment is more challenging. The results from all your class-mates have been written on the blackboard—all 22 of them. (See Table 1.1.) Summarize these results your teacher says.

Table 1.1. Heights of Students in Dr. Good's Sixth-Grade Class in Cm.

141, 156.5, 162, 159, 157, 143.5, 154, 158, 140, 142, 150, 148.5, 138.5, 161, 153, 145, 147, 158.5, 160.5, 167.5, 155, 137

---

Summarize: To describe in a straightforward, easy-to-comprehend fashion.

---

You copy the figures neatly into your notebook. You brainstorm with your teammates. Nothing. Then John speaks up—he's always interrupting in class. Shouldn't we put the heights in order from smallest to largest? "Of course," says the teacher, "you should always begin by ordering your observations." This would be an excellent exercise for the reader as well, but if you are impatient, take a look at Table 1.2.

Table 1.2. Heights of Students in Dr. Good's Sixth-Grade Class in Cm., Ordered from Shortest to Tallest

137.0  138.5  140.0  141.0  142.0  143.5  145.0  147.0  148.5  150.0  153.0
154.0  155.0  156.5  157.0  158.0  158.5  159.0  160.5  161.0  162.0  167.5

"I know what the minimum is," you say—come to think of it, you are always blurting out in class, too—"137 millimeters, that's Tony."

"The maximum, 167.5, that's Pedro, he's tall," hollers someone from the back of the room. As for the median height, the one in the middle is just 153 centime-ters (or is it 154)?

## 1.3   Picturing Data

My students at St. John's weren't through with their assignments. It was impor-tant for them to build on and review what they'd learned in the fifth grade, so I had them draw pictures of their data. Not only is drawing a picture fun, but pictures

---

[2]A hint for self-study, especially for the professional whose college years are long behind him or her: Write down and maintain a list of any new terms you come across in this text along with an example of each term's use.

and graphs are an essential first step toward recognizing patterns. In the next few sections and again in Chapter 2, we consider a variety of different ways you can visualize your data.

## 1.3.1 Graphs

You're probably already familiar with the pie chart, bar chart, and cumulative frequency distribution. Figure 1.1, 1.2, and 1.4, prepared with the aid of Stata® are examples.[3] The box and whiskers plot, Figure 1.3, and the scatter plot, Figure 1.6, may be new to you.

The pie chart illustrates percentages; the area of each piece corresponds to the relative frequency of that group in the sample. In Figure 1.1, we compare the ethnicity of students at St. Johns and a nearby public school.

The bar chart (Figure 1.2) illustrates frequencies, the height of each bar corresponding to the frequency or number in each group. Common applications of the bar chart include comparing the GNP's of various countries, and the monthly sales figures of competing salespersons.

The box and whiskers plot (Figure 1.3) provides the minimum, median, and maximum of each group side-by-side in chart form for comparison. The box extends over the interquartile range from the 25th to the 75th percentile. The line in the middle is the median. The whiskers reach out to the minimum and maximum of each sample. This particular plot allows us to make a side-by-side comparison of the heights of the two groups that make up my classroom. (Note that I had more girls than guys in my class and that in this age group, the girls tend to be taller, although both the shortest and the tallest students are boys.)

The cumulative frequency distribution (Figures 1.4 and 1.5) starts at 0 or 0%, goes up one notch for each individual counted, and ends at 1 or 100%. A frequent application, part of the modeling process, is to compare a theoretical distribution with the distribution that was actually observed.

Sixth graders make their charts by hand. You and I can let our computers do the work, either via a spreadsheet like Excel® or Quarto® or with a statistics program like Stata® or SPSS®.

Figure 1.6, a scatter plot, allows us to display the values of two observations on the same person simultaneously. Again, note how the girls in the sixth grade form a taller, more homogenous group than the boys. To obtain the data for this figure, I had my class get out their tape measures a second time and rule off the distance from the fingertips of the left hand to the fingertips of the right while the student they were measuring stood with arms outstretched like a big bird. After the assistant principal had come and gone (something about how the class was a little noisy, and although we were obviously having a good time, could we just be a little quieter), we recorded our results.

These plots and charts have several purposes. One is to summarize the data. Another is to compare different samples or different populations (girls versus

[3]Details of how to obtain copies of Stata and other statistics packages cited in this text may be found in Appendix 1.

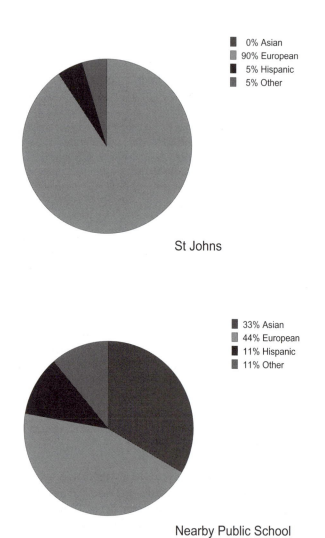

**Figure 1.1.** Students in Dr. Good's sixth-grade class. Pie chart. Distribution by ethnicity; the area of each shaded area corresponds to the relative frequency of that group.

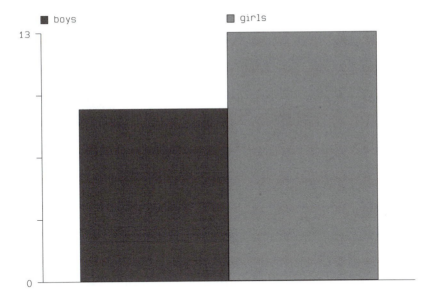

**Figure 1.2.** Students in Dr. Good's sixth-grade class. Bar chart. Distribution by sex; the height of each bar corresponds to the frequency of that group in the population.

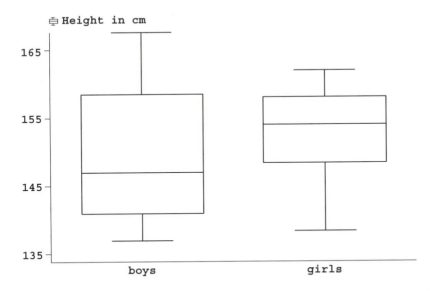

**Figure 1.3.** Students in Dr. Good's sixth-grade class. Box and whisker plot of student heights, by sex. Each box extends between the 25th and 75th percentile of the corresponding group. The line in the center of each box points to the median. The ends of the whiskers highlight the minimum and the maximum. The width of each box corresponds to the frequency of that group in the population.

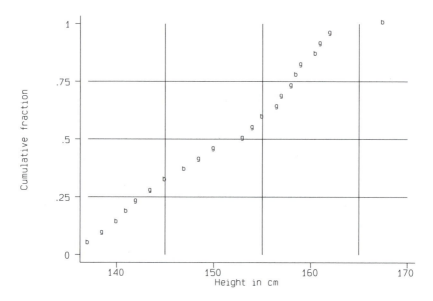

**Figure 1.4.** Students in Dr. Good's sixth-grade class. Cumulative frequency distribution of heights. The symbols b and g denote a boy and a girl, respectively.

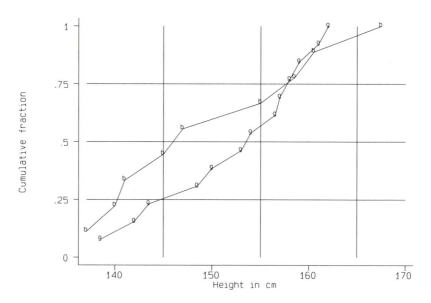

**Figure 1.5.** Students in Dr. Good's sixth-grade class. Cumulative frequency distribution of heights by sex. Symbols b and g denote a boy and a girl, respectively.

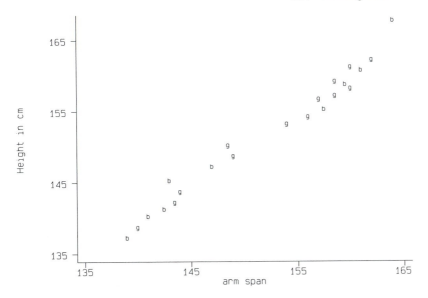

**Figure 1.6.** Students in Dr. Good's sixth-grade class. Scatter plot of arm span in cm. versus height in cm., by sex. The symbols b and g denote a boy and a girl, respectively.

boys, my class versus your class). But their primary value is as an aid to critical thinking. The figures in this specific example may make you start wondering about the uneven way adolescents go about their growth. The exciting thing, whether you are a parent or a middle-school teacher, is to observe how adolescents get more heterogeneous, more individual with each passing year.

---

**Begin with a Picture**

Charts and graphs have three purposes:

1. to summarize a set of observations
2. to compare two sets of observations, (boys vs girls, U.S. vs CA)
3. to stimulate our thinking.

---

## 1.3.2   From Observations to Questions

You may want to formulate your theories and suspicions in the form of questions: Are girls in the sixth-grade taller on the average than sixth-grade boys (not just those in Dr. Good's sixth-grade class)? Are they more homogeneous in terms of height? What is the average height of a sixth grader? How reliable is this estimate? Can height be used to predict arm span in sixth grade? at all ages?

You'll find straightforward techniques in subsequent chapters for answering these and other questions using resampling methods. First, I suspect, you'd like the answer to one really big question: Is statistics really much more difficult than the sixth-grade exercise we just completed? No, this is about as complicated as it gets.

Symbols are used throughout this book, so it helps if you've some experience with algebraic formula. But you can make it successfully and enjoyably through the first seven chapters of this text, even if you've never taken a college math course. If you're already knowledgeable in some academic discipline such as biology, or physics, or sociology, you can apply what you're reading immediately. In fact, the best way to review and understand statistics is to take some data you've collected and/or borrowed from one of your textbooks and apply the techniques in this book. Do this now. Take a data set containing 20 or 30 observations. Find the minimum, median, and maximum. Make some charts, frequency distributions, and scatter plots. If you have kids or a younger sibling, pull a Tom Sawyer and see if you can get them to do it. Otherwise, get out your favorite spreadsheet or statistics package and start working through the tutorials.

## 1.3.3  Contingency Tables

We can and should use tables to communicate our results. Hint: less is better.

Although St. John's is an Episcopalian school, my students came from a wide variety of religious backgrounds as the following *contingency table* reveals:

| Episcopalian | Protestant | Catholic | Jewish | Other |
|---|---|---|---|---|
| 6 | 5 | 6 | 2 | 3 |

We can use a similar sort of table to record the responses of my students to a poll made of their reaction to our new team approach. Did they like to work in teams of three?

| Strongly Dislike | Dislike | O.K. | Like | Like a Lot |
|---|---|---|---|---|
| 2 | 3 | 5 | 5 | 7 |

If we group the heights of my students into intervals or categories as in Table 1.3, we can also present the heights in a special form of bar chart known as a *histogram* (Figure 1.7). In a histogram, the height of each vertical bar corresponds to the frequency or number in that ordered category.

Table 1.3. Student Height in Centimeters

| 137.5 | 142.5 | 147.5 | 152.5 | 157.5 | 162.5 | 167.5 |
|---|---|---|---|---|---|---|
| 2 | 4 | 3 | 3 | 6 | 3 | 1 |

How many categories? Freedman and Diaconis [1981] suggest we use an interval of length $2(IQR)n^{-1/3}$, where $IQR$ stands for the interquartile range (see

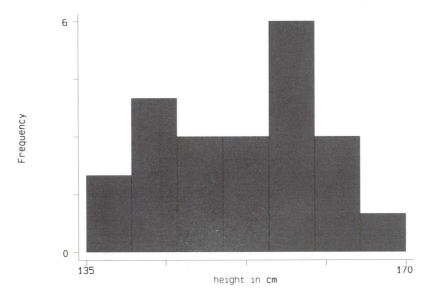

**Figure 1.7.** Histogram of heights in Dr. Good's class.

Section 1.3.1), and $n$ is the sample size. For my class of $n = 22$ students with $IQR = 15$, the recommended length is approximately 10 cm, twice the length of the interval we used.[4]

Or we could communicate this same information in the form of a one-way scatter plot (Figure 1.8), allowing us to quickly compare the boys and the girls.

## 1.3.4   Types of Data

Some variables we can observe and measure precisely to the nearest millimeter or even the nearest nanometer (billionth of a meter) if we employ a laser measuring device. We call such data *continuous*. Such data is commonly recorded on a *metric* scale such that 4 is twice 2 is twice 1. Some data, like that of the preceding section, fall into broad not-necessarily-comparable *categories*: males and females or white, black, and Hispanic. And some observations fall in between: they can be *ordered*, but they can't be added or subtracted the way continuous measurements can. An example would be the opinion survey in my classroom that yielded answers like "strongly favors," "favors," "undecided," "opposes," "strongly opposes."

We can compute the mode for all three types of observation, and the median for both continuous and ordered variables, but the mean only makes sense for data measured on a metric scale such as heights or test scores.[5]

---

[4]Regardless of what the formula says, we must always end up with an integral number of bins.

[5]A fourth data type, the binomial variable, which has exactly two categories, will be considered at length in the next chapter.

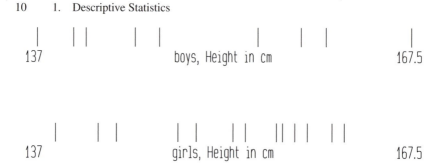

**Figure 1.8.** One-way scatter plots of student heights in Dr. Good's class.

## 1.4   Measures of Location

Far too often, we find ourselves put on the spot, forced to come up with a one-word description of our results when several pages would do. "Take all the time you like," coming from my boss, usually means, "Tell me in ten words or less."

If you were asked to use a single number to describe data you've collected, what number would you use? Probably your answer is "the one in the middle," the median that we defined earlier in this chapter. When there are an odd number of observations, it's easy to compute the *median*. The median of 2, 2, 3, 4 and 5, for example, is 3. When there are an even number of observations, the trick is to split the difference: The median of the four observations 2, 3, 4, and 5 is 3.5 halfway between the two observations in the middle.

Another answer some people would feel comfortable with is the most frequent observation or *mode*. In the sample 2, 2, 3, 4 and 5, the mode is 2. Often the mode is the same as the median or close to it. Sometimes it's quite different and sometimes, particularly when there is a mixture of populations, there may be several modes.

Consider the data on heights collected in my sixth-grade classroom, grouped as in Table 1.3. The mode is at 157.5 cm. But in this case, there may really be two modes, one corresponding to the boys, the other to the girls in the class. Often, we don't know in advance how many subpopulations there are, and the modes serve a second purpose, to help establish the number of subpopulations.

Suppose you've done a survey of college students on how they feel about nudity on television with the possible answers being 0) never, 1) rarely 2) if necessary, 3) O.K., 4) more, more. Your survey results show 6 nevers, 4 rarely's, 5 if necessary's, 3 O.K.'s and 7 more, more's. The median is "necessary," the mode appears to be "more, more." But perhaps there are really three modes, representing three populations each centered at a different response (not being a sociologist, I won't attempt to describe the differences among the "nevers," the "if necessaries," and the "more mores").

### 1.4.1   Arithmetic Mean

The mean or arithmetic average is the most commonly used measure of location today, though it can sometimes be misleading.

---

### The Center of a Population

*Median:* the value in the middle; the halfway point; that value which has equal numbers of larger and smaller elements around it.

*Arithmetic mean or arithmetic average:* the sum of all the elements divided by their number or, equivalently, that value such that the sum of the deviations of all the elements from it is zero.

*Mode:* the most frequent value. If a population consists of several subpopulations, there may be several modes.

---

The *arithmetic mean* is the sum of the observations divided by the number of observations. The mean of the sample 2,2,3,4,5, for instance, is $(2 + 2 + 3 + 4 + 5)/5$ or 3.2. The mean height of my sample of 22 sixth-grade students is $(3335.5 \text{ cm})/22 = 151.6$ cm.

The arithmetic mean is a center of gravity, in the sense that the sum of the deviations of the individual observations about it is zero. This latter property lends itself to the calculus and is the basis of many theorems in mathematical statistics.

The weakness of the arithmetic mean is that it is too easily biased by extreme values. If we eliminate Pedro from our sample of sixth graders—he's exceptionally tall for his age at 5'7"—the mean would change to $3167/21 = 150.8$ cm. The median would change to a much lesser degree, shifting from 153.5 to 153 cm.

## 1.4.2   Geometric Mean

The *geometric mean* is the appropriate measure of location when we are working with proportions or when we are expressing changes in percentages, rather than absolute values. For example, if in successive months the cost of living was 110%, 105%, 110%, 115%, 118%, 120%, 115% of the value in the base month, the average or geometric mean would be $(1.1 * 1.05 * 1.1 * 1.15 * 1.18 * 1.2 * 1.15)^{1/6}$.

## 1.5   Measures of Dispersion

As noted earlier, I don't like being put on the spot when I report my results. I'd much rather provide a picture of my data—the cumulative frequency distribution, for example, or a one-way scatter plot—than a single summary statistic. Consider the following two samples:

> Sample A:   2, 3, 4, 5, 6 (in grams),
> Sample B:   1, 3, 4, 5, 7 (in grams).

Both have the same median, 4 grams, but it is obvious the second sample is more dispersed than the first. The span or *range* of the first sample is four grams,

---

**Statistics and Their Algebraic Representation**

Suppose we have taken $n$ observations; let $x_1$ denote the first of these observations, $x_2$ the second, and so forth, down to $x_n$. Now, let us order these observations from the smallest $x_{(1)}$ to the largest $x_{(n)}$, so that $x_{(1)} \leq x_{(2)} \leq \ldots \leq x_{(n)}$.

A statistic is any single value that summarizes the characteristics of a sample. In other words, a statistic S is a function of the observations $\{x_1, x_2, \ldots, x_n\}$.

Examples include $x_{(1)}$, the minimum, and $x_{(n)}$, the maximum. If there are an odd number of observations, such as $2k + 1$, then $x_{(k+1)}$ is the median. If there are an even number of observations, such as $2k$, then the median is $\frac{1}{2}(x_{(k)} + x_{(k+1)})$.

Let $x.$ denote the sum of the observations $x_1 + x_2 + \ldots + x_n$. We may also write $x. = \sum_{i=1}^{n} x_i$. Then the arithmetic mean $\overline{x} = x./n$.

The geometric mean is equal to $(x_1 * x_2 * \ldots * x_n)^{1/n}$.

The sample variance $s^2 = \sum_{i=1}^{n}(\overline{x} - x_i)^2/(n-1)$.

The standard deviation or $L_2$-norm of the sample is the square root $s$ of the sample variance.

The first or $L_1$ sample norm is $\sum_{i=1}^{n} |\overline{x} - x_i|/(n-1)$.

---

from a minimum of 2 grams to a maximum of 6, while the values in the second sample range from 1 gram to 7; the range is six grams.

The classic measure of dispersion is the *standard deviation*, or $L_2$-norm, the square root of the sample *variance*, which is the sum of the squares of the deviations of the original observations about the sample mean, divided by the number of observations minus one. An alternative is the first or $L_1$-*norm*, the sum of the absolute values of the deviations of the original observations about the sample mean, divided by the number of observations minus one. For our first sample, the variance is $(4 + 1 + 0 + 1 + 4)/4 = 10/4$ grams$^2$, the *standard deviation* is 1.323 grams, and the $L_1$-norm is $(2 + 1 + 0 + 1 + 2)/4 = 1.5$ grams. For our second sample, the variance is $(9 + 1 + 0 + 1 + 9)/4 = 5$ grams$^2$, the standard deviation is 2.236 grams, and the $L_1$-norm is $(3 + 1 + 0 + 1 + 3)/4 = 2$ grams.

## 1.6    Sample Versus Population

Generally, though not always, we focus our attention on only a small sample of individuals drawn from a much larger population. Our objective is to make inferences about the much larger population from a detailed study of the sample.

We conduct a poll by phone of 100 individuals selected at random from a list provided by the registrar of voters, then draw conclusions as to how everyone in the county is likely to vote. We study traffic on a certain roadway five weekday mornings in a row, then draw conclusions as to what the traffic is likely to be during a similar weekday morning in the future.

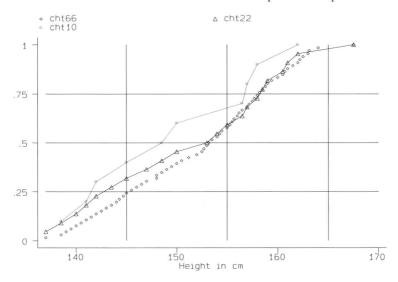

**Figure 1.9.**   Cumulative distribution functions of three samples of increasing size taken from the same population.

We study a sample rather than a population both to save time and to cut costs. Sometimes, we may have no other alternative—when we are administering a risky experimental drug, for example, or are estimating mean time to failure of a light-bulb or a disk drive.  We don't expect the sample to exactly resemble the larger population—samples often differ widely from one another just as individuals do, but our underlying philosophy is that *the larger the sample, the more closely it will resemble the population from which it is drawn.*

Figure 1.9 illustrates this point with three closely related cumulative frequency distributions.  The first, which in this instance lies above the two others, is that of a random sample of ten students taken from my sixth-grade class at St. John's School, the second, depicted earlier in Figure 1.4, that of the entire class, and the third, that of all the sixth-grade classes in the school. These three may be thought of as samples of increasing size taken from a still-larger population, such as all sixth-graders in Southern California.

Both theory and experience tell us that as the sample size increases, more and more of the samples will closely resemble the population from which they are drawn, the maxima and minima will get more extreme, while the sample median will get closer and closer to the population median.

## 1.6.1   Statistics and Parameters

A *statistic* is any single value that summarizes some aspect of a sample. A *parameter* is any single value that summarizes some aspect of an entire population. Examples of statistics include measures of location and central tendency such as the sample mode, sample median, and sample mean, extrema such as the sample minimum and maximum, and measures of variation and dispersion such as the sample standard deviation and the sample $L_1$-norm. These same measures are considered parameters when they refer to an entire population.

## 1.6.2   Estimating Population Parameters

The sample median provides us with an estimate of the population median, but it isn't the only possible estimate. We could also use the sample mean or sample mode to estimate the population median, or even the average of the minimum and the maximum. The sample median is the preferred estimate of the population median for two reasons:

It is an *unbiased* estimate of the population median, that is, if we take a very large number of samples from the population of the same size as our original sample, the median of these sample medians will be the population median.

The sample median is *consistent*, that is, if we were to take a series of successively larger samples from the population, the sample medians would come closer and closer to the population median.

In the next section, we shall consider a third property of this estimate, its precision.

## 1.6.3   Precision of an Estimate

We can seldom establish the *accuracy* of an estimate, for example, how closely the sample median comes to the unknown population median. But we may be able to establish its *precision*, that is, how closely the estimates derived from successive samples resemble one another.

The straightforward way to establish the precision of the sample median is to take a series of samples from the population, determine the median for each of these samples, and compute the standard deviation or the $L_1$-norm of these medians. Constraints on money and time make this approach unfeasible in most instances. Besides, if we had this many samples, we would combine them all into one large sample in order to obtain a much more accurate and precise estimate. One practical alternative, known as the *bootstrap*, is to treat the original sample of values as a stand-in for the population and to resample from it repeatedly, with replacement, recomputing the median each time.

In Appendix 2, we've provided the details of the bootstrap algorithm for use with C++. (Stata and S-Plus already have built-in commands.) Our algorithm is general purpose and will prove effective whether we are estimating the variance of the sample median or some other, more complicated statistic. If you don't have a computer, you could achieve the same result by writing each of the observations on a separate slip of paper, putting all the slips into an urn, and then drawing a slip at a time from the urn, replacing the slip in the urn as soon as you've recorded the number on it. You would repeat this procedure until you have a bootstrap sample of the same size as the original.

**Figure 1.10.** One way scatter plot of 50 bootstrap medians derived from a sample of heights of 22 students in Dr. Good's sixth-grade class.

As an example, our first bootstrap sample, which I've arranged in increasing order of magnitude for ease in reading, might look like this:

| | | | | | | |
|---|---|---|---|---|---|---|---|
| 138.5 | 138.5 | 140.0 | 141.0 | 141.0 | 143.5 | 145.0 | 147.0 |
| 148.5 | 150.0 | 153.0 | 154.0 | 155.0 | 156.5 | 157.0 | 158.5 |
| 159.0 | 159.0 | 159.0 | 160.5 | 161.0 | 162.0 | | |

Several of the values have been repeated as we are sampling with replacement. The minimum of this sample is 138.5, higher than that of the original sample, the maximum at 162.0 is less, while the median remains unchanged at 153.5.

| | | | | | | |
|---|---|---|---|---|---|---|---|
| 137.0 | 138.5 | 138.5 | 141.0 | 141.0 | 142.0 | 143.5 | 145.0 |
| 145.0 | 147.0 | 148.5 | 148.5 | 150.0 | 150.0 | 153.0 | 155.0 |
| 158.0 | 158.5 | 160.5 | 160.5 | 161.0 | 167.5 | | |

In this second bootstrap sample, we again find repeated values; this time the minimum, maximum, and median are 137.0, 167.5, and 148.5, respectively.

The medians of fifty bootstrapped samples drawn from our sample of sixth-graders ranged between 142.25 and 158.25 with a median of 152.75 (Figure 1.10). They provide a feel for what might have been had we sampled repeatedly from the original population.

## 1.6.4 Caveats

*The sample is not the population.* A very small or unrepresentative sample may give a completely misleading picture of the population as a whole. Take a look at the National League batting averages, particularly in the Spring. You'll find quite a number of averages over .400 and many that are close to .100. Impossible, you say: there've been no .400+ hitters since Ted Williams, and anyone with a batting average close to .100 would be on his way to the minors. But it's not impossible if the player in question has had only 10 or 20 at bats. In the long run (that is, with a larger sample of at bats) the player will start to look a lot more like we'd expect of a major leaguer.

Hospital administrators take note: If your institution only does a few of a particular procedure each year—a heart transplant, for example, don't be surprised if you have results that are much higher or lower than the national average. It may just be a matter of luck rather than any comment on your facility.

*The bootstrap will not reproduce all the relevant characteristics of a population,* only those present in the sample, and cannot be used to estimate extrema such as maximum flood height or minimum effective dose.

The bootstrap can help us to estimate the errors that result from using a sample in place of the population (see Section 5.3), but one must be cautious in its application. Of the 50 bootstrap samples taken from my sample of 22 sixth-graders, 90% of the bootstrap medians lay between 147 and 156.5. But the probability the population median lies between these values may be much larger or much smaller than 90%. Techniques for deriving more accurate interval estimates are described in Section 5.4.

## 1.7   Summary

In this chapter you learned to use summary statistics and graphs to report your findings. You learned definitions and formulas for measures of location and dispersion, and to use sample statistics to estimate the parameters of the population from which the sample was drawn. Finally, you used the bootstrap to determine the precision of your estimates.

## 1.8   To Learn More

Some of the best introductions to graphics for describing samples can be found in the manuals of such statistics packages as Stata and SAS. But see also the comprehensive reference works by Cleveland [1985, 1993] and Whittaker [1990]. Scott [1992] offers an excellent introduction to multivariate graphics.

The bootstrap has its origins in the seminal work of Jones [1956], McCarthy [1969], Hartigan [1969,1971], Simon [1969] and Efron [1979,1982]. Chapter 5 discusses its use in estimation in much more detail. One of its earliest applications to real-world data may be found in Makinodan et al [1976]. Non-technical descriptions may be found in Diaconis and Efron [1983], Efron and Tibshirani [1991], Lunneborg [1985], and Rasmussen [1987].

## 1.9   Exercises

Read over each exercise and be sure you understand what is required before you begin. Do several of the exercises using only a calculator. Use a spreadsheet or a statistics package such as Stata, S+, SAS, or StatXact to complete whatever further number of exercises you feel you need to feel comfortable with the methods you've learned. You may also download without charge, software I've developed from my home page at **http://users.oco.net/drphilgood/resamp.htm**.

Save your data on disk as we will return to these same data sets in future chapters. Make up and work through similar exercises using data that you or a colleague have collected.

1. Are the following variables categorical, ordered, or continuous? Are they measured on a metric scale? age? age in years? sex? industry rank? make of automobile? views on the death penalty? annual sales? shoe size? method of payment (cash, check, credit card)?

2. **Executive Compensation and Profitability for Seven Firms in the Aerospace Industry**

| Company | CEO Salary ($000s) | Sales ($000,000s) | Return on Equity (%) |
|---------|-----|-----|-----|
| Boeing | 846 | 19,962.0 | 11.4 |
| General Dynamics | 1041 | 11,551.0 | 19.7 |
| Lockheed | 1146 | 12,590.0 | 17.9 |
| Martin Marietta | 839 | 6,728.5 | 26.6 |
| McDonnell Douglas | 681 | 15,069.0 | 11.0 |
| Parker Hannifin | 765 | 2,437.3 | 12.4 |
| United Technology | 1148 | 19,057.1 | 13.7 |

a. How many observations in this data set?

b. Each observation consists of how many variables?

c. Compute the mean, median, minimum, and maximum of each variable.

d. What would you estimate the variance (precision) of your estimates of the mean and median to be?

e. Plot a cumulative frequency distribution for each variable.

f. Do you feel these distributions represent the aerospace industry as a whole? If not, why not?

g. Make a scatter diagram(s) for the variables.

3. Nine young Republicans were asked for their views on the following two issues:

   a. Should a woman with an unwanted pregnancy have the baby even if it means dropping out of school and going on welfare?

   b. Should the government increase welfare support for people with children?

   For both issues, 10 = strongly agree and 1 = strongly disagree. Answers are recorded in the following table:

| woman's choice | 9 | 8 | 1 | 2 | 2 | 10 | 10 | 4 | 3 |
|---------|---|---|---|---|---|----|----|---|---|
| government | 5 | 4 | 6 | 2 | 1 | 1 | 4 | 7 | 9 |

Draw a scattergram of these results.

4. Make or buy a target divided into six or more sections, and three or more concentric circles and make 50 shots at the center of the target. (If you don't have a pistol or darts, then mark off a piece of wrapping paper, put it on the floor, and drop coins on it from above.) Use a contingency table to record the number of shots, or coins, or darts in each section and each circle.

   a. How many shots (coins) would you expect to find in each section? circle?

   b. How many shots (coins) did you find?

   c. Do you feel your shooting is precise? unbiased? accurate?

**5.** Use 50–100 bootstrap samples to estimate the standard deviation and the $L_1$-norm of the mean, median, and mode of the heights of my sample of 22 sixth-graders. Which is the least precise measure of central tendency? The most precise? How do these values compare with the standard deviation and the $L_1$-norm of the original sample?

**6.**  Motrin Study 71:Site 8; Selected Variables from the Initial Physical

| Subject ID | Sex | Age | Height | Weight |
|---|---|---|---|---|
| 0118701 | M | 30 | 68 | 155 |
| 0118702 | F | 28 | 63 | 115 |
| 0118703 | F | 23 | 61 | 99 |
| 0118704 | M | 24 | 72 | 220 |
| 0118705 | M | 28 | 69 | 170 |
| 0118706 | F | 31 | 65 | 125 |
| 0118707 | M | 26 | 70 | 205 |

**a.** How many observations are there?

**b.** How many variables per observation?

**c.** Determine the minimum, median, and maximum for each variable, for each sex.

**d.** Draw scatter diagrams that will include information from all the variables.

**7.** During the 1970s, administrators at the US Public Health Service complained bitterly that while the costs of health care increased over a four-year period from $43.00 to $46.50 to $49.80 to $53.70 per patient, their budget remained the same. What was the average cost per patient over the four-year period? (Careful, this question deals with costs per patient, not total costs, and not total patients. Which measure of central tendency should you use?)

**8.**

| Country | Age | | |
|---|---|---|---|
| | 0 | 25 | 50 |
| Argentina | 65 | 46 | 24 |
| Costa Rica | 65 | 48 | 26 |
| Dominican Rep | 64 | 50 | 28 |
| Ecuador | 57 | 46 | 25 |
| El Salvador | 56 | 44 | 25 |
| Grenada | 61 | 45 | 22 |
| Honduras | 59 | 42 | 22 |
| Mexico | 59 | 44 | 24 |
| Nicaragua | 65 | 48 | 28 |
| Panama | 65 | 48 | 26 |
| Trinidad | 64 | 43 | 21 |

The preceding table records life expectancies at various ages for various Latin American countries.

**a.** Draw scatter plots for all pairs of variables.

**b.** Draw a contour plot.

9. Health care costs are still out of control in the 1990s and there is wide variation in the costs from region to region. The Community Care Network, which provides billing services for several dozen PPOs, recorded the following cost data for a single dermatological procedure, debridement:

| | | | | | | | |
|-----|-------|-------|-------|-------|-------|-------|-------|
| CIM | $198 | 200.2 | 242 | | | | |
| HAJ | 83.2 | 193.6 | | | | | |
| HAV | 197 | | | | | | |
| HVN | 93 | | | | | | |
| LAP | 105 | | | | | | |
| MBM | 158.4 | 180.4 | 160.6 | 171.6 | 170 | 176 | 187 |
| VNO | 81.2 | 103.0 | 93.8 | | | | |
| VPR | 154 | 228.8 | 180.4 | 220 | 246.4 | 289.7 | 198 | 224.4 |

**a.** Prepare both a histogram and a one-way scatter plot for this data.

**b.** Determine the median and the range.

10. In order to earn a C in Professor Good's statistics course, you should have an average score of 70 or above. To date, your test scores have read 100, 69, 65, 60, and 60. Should you get a C?

11. My Acura Integra seems to be less fuel efficient each year (or maybe its because I'm spending more time on city streets and less on freeways). Here are my miles per gallon result—when I remembered to record them—for the last three years:

| | |
|------|---|
| 1993 | 34.1 32.3 31.7 33.0 29.5 32.8 31.0 32.9 |
| 1994 | 30.2 31.4 32.2 29.9 33.3 31.4 32.0 29.8 30.6 |
| | 28.7 30.4 30.0 29.1 31.2 |
| 1995 | 30.1 30.1 29.5 28.5 29.9 30.6 29.3 32.4 30.5 30.0 |

**a.** Construct a histogram for the combined data.

**b.** Construct separate frequency distributions for each year.

**c.** Estimate the median miles per gallon for my automobile. Estimate the variance of your estimate.

12. Here are the results of a recent salary survey in the company I work for. As

with most companies, only a few make the big bucks.

| Wage | Number of employees |
|---|---|
| $10–19,000 | 39 |
| 20–39,000 | 30 |
| 40–59,000 | 14 |
| 40–49,000 | 8 |
| 50–59,000 | 5 |
| 60–69,000 | 2 |
| 70–79,000 | 1 |
| 80–89,000 | 0 |
| 90–99,000 | 1 |

**a.** Create both a histogram and a cumulative frequency diagram.

**b.** What are the mean, median, and the modal wages? Which gives you the best estimate of the average worker's take-home pay?

**c.** Which measure of central tendency would you use to estimate the costs of setting up a sister office in the northern part of your state?

13. The following vaginal titers were observed in mice 144 hours after inoculation with Herpes virus type II: 10000, 9000, 3000, 2600, 2400, 1700, 1500, 1100, 360, and 1.

**a.** What is the average value? (This is a trick question. The trick is to determine which of the averages we've discussed best represents this specific data set.)

**b.** Draw a one-way scatter plot and a histogram. (Hint: This may be difficult unless you first transform the values by taking their logarithms. Most statistics packages include this transform.)

14.

| LSAT | GPA | LSAT | GPA | LSAT | GPA | LSAT | GPA |
|---|---|---|---|---|---|---|---|
| 545 | 2.76 | 575 | 2.74 | 594 | 2.96 | 653 | 3.12 |
| 555 | 3.00 | 576 | 3.39 | 605 | 3.13 | 661 | 3.43 |
| 558 | 2.81 | 578 | 3.03 | 635 | 3.30 | 666 | 3.44 |
| 572 | 2.88 | 580 | 3.07 | 651 | 3.36 | | |

Is performance on the LSAT's related to undergraduate GPA? Draw a scattergram for the data in the preceding table. Determine the sample mean, median, and standard deviation for each of the variables. Use the bootstrap 116to evaluate the precision of your estimate of the mean.

The preceding data are actually a sample of admission scores from a total of 82 American law schools. The mean LSAT for all 82 schools is 597.55. How does this compare with your estimate based on the sample?

15. **a.** If every student in my sixth-grade class grew five inches overnight, what would the mean, median, and variance of their new heights have been?

    **b.** If I'd measured their heights in inches rather than centimeters, what would the mean, median, and variance of their heights have been? (Assume that 1 cm = 2.54 inches.)

16. Summarize the billing data from four Swedish hospitals (Hint: A picture can be worth several hundred numbers.) Be sure to create a histogram for each of the hospitals. To save you typing, you may download this data set from my home page **http://users.oco.net/drphilgood/resamp.htm**.

Hospital 1,

| 64877 | 21152 | 11753 | 1834 | 3648 | 12712 | 11914 | 14290 | 17132 |
|-------|-------|-------|------|------|-------|-------|-------|-------|
| 7030 | 23540 | 5413 | 4671 | 39212 | 7800 | 10715 | 11593 | 3585 |
| 12116 | 8287 | 14202 | 4196 | 22193 | 3554 | 3869 | 3463 | 2213 |
| 3533 | 3523 | 10938 | 17836 | 3627 | 30346 | 2673 | 3703 | 28943 |
| 8321 | 19686 | 18985 | 2243 | 4319 | 3776 | 3668 | 11542 | 14582 |
| 9230 | 7786 | 7900 | 7886 | 67042 | 7707 | 18329 | 7968 | 5806 |
| 5315 | 11088 | 6966 | 3842 | 13217 | 13153 | 8512 | 8328 | 207565 |
| 2095 | 18985 | 2143 | 7976 | 2138 | 15313 | 8262 | 9052 | 8723 |
| 4160 | 7728 | 3721 | 18541 | 7492 | 18703 | 6978 | 10613 | 15940 |
| 3964 | 10517 | 13749 | 24581 | 3465 | 11329 | 7827 | 3437 | 4587 |
| 14945 | 23701 | 61354 | 3909 | 14025 | 21370 | 4582 | 4173 | 4702 |
| 7578 | 5246 | 3437 | 10311 | 8103 | 11921 | 10858 | 14197 | 7054 |
| 4477 | 4406 | 19170 | 81327 | 4266 | 2873 | 7145 | 4018 | 13484 |
| 7044 | 2061 | 8005 | 7082 | 10117 | 2761 | 7786 | 62096 | 11879 |
| 3437 | 17186 | 18818 | 4068 | 10311 | 7284 | 10311 | 10311 | 24606 |
| 2427 | 3327 | 3756 | 3186 | 2440 | 7211 | 6874 | 26122 | 5243 |
| 4592 | 11251 | 4141 | 13630 | 4482 | 3645 | 5652 | 22058 | 15028 |
| 11932 | 3876 | 3533 | 31066 | 15607 | 8565 | 25562 | 2780 | 9840 |
| 14052 | 14780 | 7435 | 11475 | 6874 | 17438 | 1596 | 10311 | 3191 |
| 37809 | 13749 | 6874 | 6874 | 2767 | 138133 | | | |

Hospital 2

| 4724 | 3196 | 3151 | 5912 | 7895 | 19757 | 21731 | 13923 | 11859 | 8754 |
|------|------|------|------|------|-------|-------|-------|-------|------|
| 4139 | 5801 | 11004 | 3889 | 3461 | 3604 | 1855 | | | |

Hospital 3

| 4181 | 2880 | 5670 | 11620 | 8660 | 6010 | 11620 | 8600 | 12860 |
|------|------|------|-------|------|------|-------|------|-------|
| 21420 | 5510 | 12270 | 6500 | 16500 | 4930 | 10650 | 16310 | 15730 |
| 4610 | 86260 | 65220 | 3820 | 34040 | 91270 | 51450 | 16010 | 6010 |
| 15640 | 49170 | 62200 | 62640 | 5880 | 2700 | 4900 | 55820 | 9960 |
| 28130 | 34350 | 4120 | 61340 | 24220 | 31530 | 3890 | 49410 | 2820 |
| 58850 | 4100 | 3020 | 5280 | 3160 | 64710 | 25070 | | |

Hospital 4

| | | | | | | | | | |
|---|---|---|---|---|---|---|---|---|---|
| 10630 | 81610 | 7760 | 20770 | 10460 | 13580 | 26530 | 6770 | 10790 |
| 8660 | 21740 | 14520 | 16120 | 16550 | 13800 | 18420 | 3780 | 9570 |
| 6420 | 80410 | 25330 | 41790 | 2970 | 15720 | 10460 | 10170 | 5330 |
| 10400 | 34590 | 3380 | 3770 | 28070 | 11010 | 19550 | 34830 | 4400 |
| 14070 | 10220 | 15320 | 8510 | 10850 | 47160 | 54930 | 9800 | 7010 |
| 8320 | 13660 | 5850 | 18660 | 13030 | 33190 | 52700 | 24600 | 5180 |
| 5320 | 6710 | 12180 | 4400 | 8650 | 15930 | 6880 | 5430 | 6020 |
| 4320 | 4080 | 18240 | 3920 | 15920 | 5940 | 5310 | 17260 | 36370 |
| 5510 | 12910 | 6520 | 5440 | 8600 | 10960 | 5190 | 8560 | 4050 |
| 2930 | 3810 | 13910 | 8080 | 5480 | 6760 | 2800 | 13980 | 3720 |
| 17360 | 3770 | 8250 | 9130 | 2730 | 18880 | 20810 | 24950 | 15710 |
| 5280 | 3070 | 5850 | 2580 | 5010 | 5460 | 10530 | 3040 | 5320 |
| 2150 | 12750 | 7520 | 8220 | 6900 | 5400 | 3550 | 2640 | 4110 |
| 7890 | 2510 | 3550 | 2690 | 3370 | 5830 | 21690 | 3170 | 15360 |
| 21710 | 8080 | 5240 | 2620 | 5140 | 6670 | 13730 | 13060 | 7750 |
| 2620 | 5750 | 3190 | 2600 | 12520 | 5240 | 10260 | 5330 | 10660 |
| 5490 | 4140 | 8070 | 2690 | 5280 | 18250 | 4220 | 8860 | 8200 |
| 2630 | 6560 | 9060 | 5270 | 5850 | 39360 | 5130 | 6870 | 18870 |
| 8260 | 11870 | 9530 | 9250 | 361670 | 2660 | 3880 | 5890 | 5560 |
| 7650 | 7490 | 5310 | 7130 | 5920 | 2620 | 6230 | 12640 | 6500 |
| 3060 | 2980 | 5150 | 5270 | 16600 | 5880 | 3000 | 6140 | 6790 |
| 6830 | 5280 | 29830 | 5320 | 7420 | 2940 | 7730 | 11630 | 9480 |
| 16240 | 2770 | 6010 | 4410 | 3830 | 3280 | 2620 | 12240 | 4120 |
| 5030 | 8010 | 5280 | 4250 | 2770 | 5500 | 7910 | 2830 | 11940 |
| 9060 | 20130 | 10150 | 6850 | 10160 | 7970 | 12960 | 3155 | |

17. Discuss the various reasons why one might take a sample, rather than examine an entire population. How might we tell when a sample is large enough?

# CHAPTER 2

# Testing a Hypothesis

In Chapter 1, we learned how to describe a sample and how to use the sample to describe and estimate the parameters of the population from which it was drawn. In this chapter, we learn how to frame hypotheses and alternatives about the population and to test them using the relabeling method. We learn the fundamentals of probability and apply them in practical situations. And we learn how to draw random, representative samples that can be used for testing and estimation.

## 2.1  A Laboratory Experiment

Shortly after I received my doctorate in statistics,[1] I decided that if I really wanted to help bench scientists apply statistics, I ought to become a scientist myself. So back to school[2] I went to learn all about physiology and aging in cells raised in Petri dishes.

I soon learned there was a great deal more to an experiment than the random assignment of subjects to treatments. In general, 90% of the effort was spent in mastering various arcane laboratory techniques, 9% in developing new techniques to span the gap between what had been done and what I really wanted to do, and a mere 1% on the experiment itself. But the moment of truth came finally—it had to if I were to publish and not perish—and I succeeded in cloning human diploid fibroblasts in eight culture dishes: Four of these dishes were filled with a conventional nutrient solution and four held an experimental "life-extending" solution to which vitamin E had been added.

I waited three weeks with my fingers crossed—there is always a risk of contamination with cell cultures—but at the end of this test period three dishes of each type had survived. My technician and I transplanted the cells, let them grow

---

[1] From the University of California at Berkeley.

[2] The Wistar Institute, Philadelphia, PA, and the W. Alton Jones Cell Science Center in Lake Placid, NY.

for 24 hours in contact with a radioactive label, and then fixed and stained them before covering them with a photographic emulsion.

Ten days passed and we were ready to examine the autoradiographs. Two years had elapsed since I first envisioned this experiment and now the results were in: I had the six numbers I needed.

"I've lost the labels," my technician said as she handed me the results.

"What!?" Without the labels, I had no way of knowing which cell cultures had been treated with vitamin E and which had not.

"121, 118, 110, 34, 12, 22." I read and reread these six numbers over and over again. If the first three counts were from treated colonies and the last three were from untreated, then I had found the fountain of youth. Otherwise, I really had nothing to report.

## 2.2   Analyzing the Experiment

How had I reached this conclusion? Let's take a second, more searching look at the problem of the missing labels. First, we identify the hypothesis and the alternative(s) of interest.

I wanted to assess the life-extending properties of a new experimental treatment with vitamin E. To do this, I had divided my cell cultures into two groups: one grown in a standard medium and one grown in a medium containing vitamin E. At the conclusion of the experiment and after the elimination of several contaminated cultures, both groups consisted of three independently treated dishes.

My *null hypothesis* was that the growth potential of a culture would not be affected by the presence of vitamin E in the media: All the cultures would have equal growth potential. The *alternative* of interest was that cells grown in the presence of vitamin E would be capable of many more cell divisions

Under the null hypothesis, the labels "treated" and "untreated" provide no information about the outcomes: The observations are expected to have more or less the same values in each of the two experimental groups. If they do differ it should only be as a result of some uncontrollable random fluctuation. Thus, if this null or no-difference hypothesis were true, I was free to exchange the labels.

The next step in the permutation method is to choose a test statistic that discriminates between the hypothesis and the alternative. The statistic I chose was the sum of the counts in the group treated with vitamin E. If the alternative is true, most of the time this sum ought to be larger than the sum of the counts in the untreated group. If the null hypothesis is true, that is, if it doesn't make any difference which treatment the cells receive, then the sums of the two groups of observations should be approximately the same. One sum might be smaller or larger than the other by chance, but most of the time the two shouldn't be all that different.

The third step in the permutation method is to compute the test statistic for each of the possible relabelings and compare these values with the value of the test statistic as the data was labeled originally. Fortunately, I'd kept a record of the treatments independent of my technician. In fact, I'd deliberately not let my technician know which cultures were which in order to ensure she would give them equal care in handling. As it happened, the first three observations my technician showed me—121, 118, and 110 were those belonging to the cultures that received vitamin E. The value of the test statistic for the observations as originally labeled is $349 = 121 + 118 + 110$.

I began to rearrange (permute) the observations, randomly reassigning the six labels, three "treated" and three "untreated," to the six observations. For example: treated, 121 118 34, and untreated, 110 12 22. In this particular rearrangement, the sum of the observations in the first (treated) group is 273. I repeated this step till all $\binom{6}{3} = 20$ distinct rearrangements had been examined.[3]

|     | First Group | | | Second Group | | | Sum of First Group |
| --- | --- | --- | --- | --- | --- | --- | --- |
| 1.  | 121 | 118 | 110 | 34  | 22  | 12  | 349 |
| 2.  | 121 | 118 | 34  | 110 | 22  | 12  | 273 |
| 3.  | 121 | 110 | 34  | 118 | 22  | 12  | 265 |
| 4.  | 118 | 110 | 34  | 121 | 22  | 12  | 262 |
| 5.  | 121 | 118 | 22  | 110 | 34  | 12  | 261 |
| 6.  | 121 | 110 | 22  | 118 | 34  | 12  | 253 |
| 7.  | 121 | 118 | 12  | 110 | 34  | 22  | 251 |
| 8.  | 118 | 110 | 22  | 121 | 34  | 12  | 250 |
| 9.  | 121 | 110 | 12  | 118 | 34  | 22  | 243 |
| 10. | 118 | 110 | 12  | 121 | 34  | 22  | 240 |
| 11. | 121 | 34  | 22  | 118 | 110 | 12  | 177 |
| 12. | 118 | 34  | 22  | 121 | 110 | 12  | 174 |
| 13. | 121 | 34  | 12  | 118 | 110 | 22  | 167 |
| 14. | 110 | 34  | 22  | 121 | 118 | 12  | 166 |
| 15. | 118 | 34  | 12  | 121 | 110 | 22  | 164 |
| 16. | 110 | 34  | 12  | 121 | 118 | 22  | 156 |
| 17. | 121 | 22  | 12  | 118 | 110 | 34  | 155 |
| 18. | 118 | 22  | 12  | 121 | 110 | 34  | 152 |
| 19. | 110 | 22  | 12  | 121 | 118 | 34  | 144 |
| 20. | 34  | 22  | 12  | 121 | 118 | 110 | 68  |

The sum of the observations in the original vitamin-E-treated group, 349, is equaled only once and never exceeded in the 20 distinct random relabelings. If chance alone is operating, then such an extreme value is a rare, only a 1-time-in-20 event. If I reject the null hypothesis and embrace the alternative that the treatment is effective and responsible for the observed difference, I only risk making an error and rejecting a true hypothesis one in every 20 times.

In this instance, I did make just such an error. I was never able to replicate the observed life-promoting properties of vitamin E in other repetitions of this

---

[3]Determination of the number of relabelings, "6 choose 3" in the present case, is considered in Section 2.5.1

experiment. Good statistical methods can reduce and contain the probability of making a bad decision, but they cannot eliminate the possibility.

## 2.3    Some Statistical Considerations

The preceding experiment, simple though its analysis may be, raises a number of concerns that are specifically statistical in nature. How do we go about framing a hypothesis? If we reject a hypothesis, what are the alternatives? Why must our conclusions always be in terms of probabilities rather than certainties? We consider each of these issues in turn.

### 2.3.1    Framing the Hypothesis

The hypothesis that we tested in this example, that cell cultures raised in medium to which vitamin E has been added have no particular advantage in growth potential over cultures raised in ordinary medium, is what statisticians term a *null hypothesis*. It is based on the assumption the various samples tested are all drawn from the same hypothetical population, or, at least, from populations for which the variable we observe all have the same distribution. Thus, the null hypothesis that the distribution of the heights of sixth-grade boys is the same as the distribution of the heights of sixth-grade girls is false, while the null hypothesis that distribution of the heights of sixth-grade boys in Orange County is the same as the distribution of the heights of sixth-grade boys in neighboring Los Angeles County is probably true.

The majority of hypotheses considered in this text are null hypothesis, but they are not the only possible hypotheses. For example, we might want to test whether all cultures treated with vitamin E would show an increase of at least three generations in life span. As it is difficult to test such a hypothesis, we would probably perform an initial *transformation* of the data, subtracting three from each of the observations in the vitamin-E-treated sample. Then, we would test the null hypothesis that there is no difference between the observations from the original untreated sample and the transformed observations.[4]

### 2.3.2    Hypothesis Versus Alternative

Whenever we reject a hypothesis, we accept an *alternative*.[5] In the present example, when we rejected the null hypothesis, we accepted the alternative that vitamin E had a beneficial effect. Such a test is termed *one-sided*. A two-sided test would also guard against the possibility that vitamin E had a detrimental effect. A two-sided test would reject for both extremely large and extremely small values of our test statistic. Such a test will be considered in the next chapter. In more complex

---

[4]See Exercise 5 at the end of this chapter.
[5]Or we may conclude we need more data before we can reach a decision.

situations, such as those considered in Chapter 8, there can be many different tests depending on which alternatives are of greatest interest.

### 2.3.3  Unpredictable Variation

In reviewing the possible outcomes of this experiment, the phrase "most of the time" was used repeatedly. As we saw in Chapter 1, something that is "predominantly" true may still be false on occasion. We expect a *distribution* of outcomes. Vitamin E indeed may have a beneficial effect, but this effect may not affect all cultures equally. Too many other factors can intervene. Any conclusions we draw must be probabilistic in nature. Always, we risk making an error and rejecting a true hypothesis some percentage of the time.[6]

## 2.4   Two Types of Populations

In Chapter 1, we studied samples taken from real accessible populations, that is, populations of actual individuals we could reach out and touch. But in many scientific studies, the populations we examine are only hypothetical. Examples include potential sixth-grade students, women 25 to 40 years of age who might someday take aspirin, the possible outcomes of 25 successive coins tosses. Our choice of words—potential, possible, might be—confirms the distinction. The good news is that we are able to characterize these hypothetical populations much the same way we characterize the accessible ones, by specifying means, medians, standard deviations, and percentiles.

The reason we need to consider hypothetical populations is that all our observations have a *stochastic* or random component. In some instances, as when we repeatedly measure a student's height with a ruler, the random component is inherent in the measuring process, so-called observation error; in others, the variation is part of the process itself. What will be the effect of my taking an aspirin? The answer depends on my current state of health, my current body temperature, even the time of day. Broaden the scope of this question, ask, for example, what will be the effect of a single aspirin on an individual selected at random, and we introduce a dozen more sources of variation—such as age and sex.

### 2.4.1  Predicting the Unpredictable

Many of these or similar observations we make in practice seem to have so many interacting causes as to be entirely unpredictable. Examples include the outcome of a flip of a coin or the roll of a die, the response to a medication, or the change in price of a commodity between yesterday and today. The possible outcomes may be equally likely as in the flip of a coin or weighted in favor of a specific

---

[6]A possible exception might arise when we have hundreds of thousands of observations. There might be a chance even then, albeit a very small one, of our making an error.

outcome as in "Take two aspirins and you'll probably feel better in the morning." Even in this latter case, some patients won't get better, while others will acquire additional symptoms resulting directly from the medication.

Although we may not be able to predict the outcome of any individual trial, still we may be able to say something about the outcome of a series of trials. In the next section, we consider the simplest of such trials, ones with exactly two possible outcomes.

## 2.5    Binomial Outcomes

The simplest stochastic variable is the binomial trial—success or failure, agree or disagree, heads or tails. If the coin is fair, that is, the only difference between the two outcomes lies in their names, then the probability of throwing a head is 1/2, and the probability of throwing a tail is also 1/2.

By definition, the probability that something will happen is 1, the probability that nothing will occur is 0. All other probabilities are somewhere in between.[7]

What about the probability of throwing heads twice in a row? Ten times in a row? If the coin is fair and the throws independent of one another, the answers are easy: $1/4^{th}$ and $1/1024^{th}$ or $(1/2)^{10}$.

These answers are based on our belief that when the only differences among several possible outcomes are their labels, "heads" and "tails," for example, the various outcomes will be equally likely. If we flip two fair coins or one fair coin twice in a row, there are four possible outcomes HH, HT, TH, and TT. Each outcome has equal probability of occurring. The probability of observing the one outcome in which we are interested is 1 in 4 or 1/4th. Flip the coin ten times and there are $2^{10}$ or a thousand possible outcomes; one such outcome might be described as HTTTTTTTTH.

Unscrupulous gamblers have weighted coins so that heads comes up more often than tails. In such a case, there is a real difference between the two sides of the coin and the probabilities will be different from those described above. Suppose as a result of weighting the coin, the probability of getting a head is now $p$, where $0 \le p \le 1$, and the complementary probability of getting a tail (or not getting a head) is $1 - p$, because $p + 1 - p = 1$. Again, we ask the question, what is the probability of getting two heads in a row? The answer is $p^2$. Here is why:

To get two heads in a row, we must throw a head on the first toss, which we expect to do in a proportion $p$ of attempts. Of this proportion, only a further fraction $p$ of two successive tosses also end with a head, that is, only $p * p$ trials result in HH. Similarly, the probability of throwing ten heads in a row is $p^{10}$.

By the same line of reasoning, we can show the probability of throwing nine heads in a row followed by a tail when we use the same weighted coin each time

---

[7]If you want to be precise, the probability of throwing a head is probably only 0.49999, and the probability of a tail is also only 0.49999. The left over probability of 0.00002 is the probability of all the other outcomes—the coin stands on edge, a sea gull drops down out of the sky and takes off with it, and so forth.

is $p^9(1 - p)$. What is the probability of throwing 9 heads in 10 trials? Is it also $p^9(1 - p)$? No, for the outcome "nine heads out of ten" includes the case where the first trial is a tail and all the rest are heads, the second trial is a tail and all the rest are heads, the third trial is ..., and so forth. Ten different ways in all. These different ways are *mutually exclusive*, that is, if one of these events occurs, the others are excluded. The probability of the overall event is the sum of the individual probabilities or $10p^9(1 - p)$.

What is the probability of throwing exactly 5 heads in 10 tosses of a coin? The answer to this last question requires we understand something about permutations and combinations, a concept that will be extremely important in succeeding chapters.

---

**In the Long Run: Some Misconceptions**

When events occur as a result of chance alone, anything can happen and usually will. You roll craps seven times in a row, or you flip a coin ten times and ten times it comes up heads. Both these events are unlikely, but they are not impossible. Before reading the balance of this section, test yourself by seeing if you can answer the following :

You've been studying a certain roulette wheel that is divided into 38 sections for over four hours now and not once during those four hours of continuous play has the ball fallen into the number 6 slot. You conclude that

  (1) number 6 is bound to come up soon and bet on it;
  (2) the wheel is fixed so that number 6 will never come up;
  (3) the odds are exactly what they've always been and in the next four hours number 6 will probably come up about 1/38th of the time.

If you answered (2) or (3) you're on the right track. If you answered (1), think about the following equivalent question:

You've been studying a series of patients treated with a new experimental drug all of whom died in excruciating agony despite the treatment. Do you conclude the drug is bound to cure somebody sooner or later and take it yourself when you come down with the symptoms? Or do you decide to abandon this drug and look for an alternative?

---

## 2.5.1  Permutations and Rearrangements

Suppose, we have three horses in a race. Call them A, B, and C. A could come in first, B could come in second, and C would be last. ABC is one possible outcome or permutation. But so are ACB, BAC, BCA, CAB, CBA. Six possibilities (permutations) in all. Now suppose we have a nine-horse race. We could write down all the possibilities, or we could use the following trick: We choose a winner (nine possibilities); we choose a second place finisher (eight remaining possibilities), and so forth until all positions are assigned. A total of

$9! = 9 \times 8 \times 7 \times 6 \times 5 \times 4 \times 3 \times 2 \times 1$ possibilities in all. Had there been $N$ horses in the race, there would have been $N!$ possibilities.

Normally in a horse race, all our attention is focused on the first three finishers. How many possibilities are there? Using the same reasoning, it is easy to see there are $9 \times 8 \times 7$ possibilities or $9!/6!$. Had there been $N$ horses in the race, there would have been $N!/(N-3)!$ possibilities.

Suppose we ask a slightly different question: In how many different ways can we select three horses from nine entries without regard to order (that is, we don't care which comes first, which second, or which third). In the previous example, we distinguished between first, second, and third place finishers; now we're saying the order of finish doesn't make any difference. We already know there are $3! = 3 \times 2 \times 1 = 6$ different permutations of the three horses that finish in the first three places. So we take our answer to the preceding question $9!/6!$ and divide this answer in turn by $3!$. We write the result as $\binom{9}{3}$ which is usually read as 9 choose 3. Note that $\binom{9}{3} = \binom{9}{6}$.

In how many different ways can we assign nine cell cultures to two unequal experimental groups, one with three cultures and one with six? This would be the case if we had nine labels and three of the labels read "vitamin E" while six read "controls." If we could distinguish the individual labels, we could assign them in $9!$ different ways. But the order they are assigned in each of the experimental groups, $3!$ ways in the first instance and $6!$ in the other, won't affect the results. Thus there are only $9!/6!3!$ or $\binom{9}{3} = 84$ distinguishable ways. We can generalize this result to show the number of distinguishable ways $N$ items can be assigned to two groups, one of $k$ items, and one of $N-k$ is $\frac{N!}{k!(N-k)!}$.

What if we were to divide these same nine cultures among three equal-sized experimental groups? Then we would have $9!/3!3!3!$ distinguishable ways, written as $\binom{9}{3\ 3\ 3}$.

### 2.5.2   Back to the Binomial

We used horses in this example, but the same reasoning can be applied to coins or survivors in a clinical trial.[8] What is the probability of 5 heads in 10 tosses? What is the probability that 5 of 10 breast cancer patients will still be alive after six months?

We answer this question in two stages. First, what is the number of different ways we can get five heads in 10 tosses? We could have thrown HHHHHTTTTT or HHHHTHTTTT, or some other combination of five heads and five tails for a total of 10 choose 5 or $10!/(5!5!)$ ways. The probability the first of these events occurs—five heads followed by five tails—is $(1/2)^{10}$. Combining these results

---

[8]If, that is, the probability of survival is the same for every patient. When there are obvious differences from trial to trial—one is an otherwise healthy 35-year-old male, the other an elderly 89-year-old who has just recovered from pneumonia—this simple binomial model would not apply. See also Section 2.7.1.

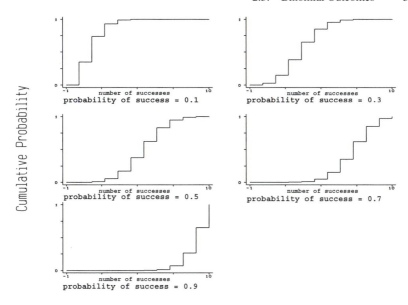

**Figure 2.1.** Cumulative probability distribution for the binomial with 10 trials for various values of $p$ from 0.1 to 0.9.

yields

$$\Pr\{5 \text{ heads in } 10 \text{ throws of a fair coin}\} = \binom{10}{5}(1/2)^{10}.$$

We can generalize the preceding to an arbitrary probability of success $p$, $0 \leq p \leq 1$. The probability of failure is $1 - p$. The probability of $k$ successes in $n$ trials is given by the binomial formula

$$\binom{n}{k}(p)^k(1 - p)^{n-k} \qquad \text{for } 0 \leq k \leq n.$$

Figure 2.1 depicts the probability of $k$ or fewer successes in 10 trials for various values of $p$. The expected number of successes in many repetitions of $n$ binomial trials each of which has a probability $p$ of success is $np$; the expected variance is $np(1 - p)$.

### 2.5.3  Probability

We have already discussed probability in an informal way, and the purpose of this section is merely to summarize our previous discussion. Let $\Pr\{A\}$ denote the probability that an event $A$ occurs, for example, a toss of a coin yields heads, or an advertisement for your product is read by at least 1000 people, or a patient who takes two aspirins feels better within the hour.

For any event $A$, $0 \leq \Pr\{A\} \leq 1$. An impossible event has probability 0; a sure thing has probability 1. Most of the other rules of probability are equally obvious:

Either something happens or it doesn't: $\Pr\{A\}+\Pr\{\text{not }A\} = 1$. The probability that a fair coin comes up heads is 1/2; the probability it comes up tails is also 1/2. More generally, if we have $k$ *equally likely* mutually exclusive events, that is, the only distinction among them is their labels, then each will occur with probability $1/k$. Consider, for example, the probability of throwing a 6 with a six-sided die or that "Whistler's Brother" wins, when "Brother" is one of 10 horses in a race where each animal has been perfectly handicapped by the addition of weights. In this latter case, a bet on "Whistler's Brother," at less than $10 - 1$ odds is an indication you feel you are a better handicapper than the person who specified the weights.

If $B$ is a collection of events that includes $A$, then $\Pr\{B\} \geq \Pr\{A\}$.

If $C$ is a collection of events $\{B_1, B_2, \ldots, B_n\}$ that are mutually exclusive, then $\Pr\{C\} = \Pr\{B_1\} + \Pr\{B_2\} + \ldots + \Pr\{B_n\} = \Sigma\,\Pr\{B_i\}$. An example of such a collection is the event throws heads at least four out of five times in a row which is made up of two mutually exclusive events $B_1 = \{$throws heads four out of five times in a row$\}$ and $B_2 = \{$throws heads five out of five times in a row$\}$. These events are mutually exclusive because only one of them can occur. The latter consists of the single event HHHHH, but the former is actually composed of five mutually exclusive events HHHHT, HHHTH, HHTHH, HTHHH, and THHHH. As there are $2^5 = 32$ distinct possibilities in all, $\Pr\{B_2\} = 1/32$, $\Pr\{B_1\} = 5/32$, and $\Pr\{C\} = 6/32$.

Actually, each one of these so-called fundamental events is made up of still more detailed possibilities, such as "the first time I threw the coin in the air, it spun around five times, was touched briefly by a wind current, bounced an inch above by my hand, bounced a second time a mere fraction of an inch, and then came to rest head upwards...," but you get the idea.

## 2.6   Independence

Underlying most statistical procedures is the assumption that successive observations are independent of one another. The sequence of binomial trials described above is just one example. Violate this assumption and resampling methods such as the bootstrap and permutation tests are no longer applicable.

Independence means all of the following:

1. Independent trials do not influence each other's outcome.

2. If we know the outcome of one of the trials, it won't provide us with additional insight into the possible outcome of the other trials.

3. The probability of the joint outcome of two independent trials can be written as the product of the probabilities of the individual trials. (We took advantage of this multiplication rule in Section 2.5)

When events are both independent and identically distributed, we are free to exchange the labels without affecting our interpretation of the results.

An example of independent events is $A$: the coin came up heads, $B$: the person residing at 1063 Elm Street votes Republican. The best way to understand the concept of independence is to consider a few other simple examples. You flip 10 coins once each. Are the 10 outcomes independent? Of course. You ask the first 10 people you meet in the street if they are against compulsory pregnancy; are their answers independent? Maybe. And maybe the 10 just walked out of a Planned Parenthood meeting.

What about shuffling cards? Shuffle the cards thoroughly (at least seven standard riffle shuffles is recommended) and successive deals are independent. Shuffle them carelessly and the beginning of one card game will bear a strong resemblance to the finish of the last.

Measure the blood pressures of two different individuals and the values are independent, right? Maybe: If the measurements were taken by an experienced individual in a quiet room. But walk into a physiology class some day when the class is just learning to take blood pressure. Everybody is hyper and all the measurements tend to be on the high side.

What about two successive measurements taken on the same individual such as blood pressures or test scores? Obviously these measurements are dependent. But what about the first observation and the difference between it and the next measurement? These two may be independent. If the Dow-Jones Index is high today, we're probably safe in believing it will be on the high side tomorrow. But we don't know whether it will go up or down in the interval or by how many points.

## 2.7  Selecting the Sample

To evaluate your tentative theory, you plan to draw a sample from a much larger population, then apply resampling methods to estimate the characteristics of the population from which the sample is derived. But will the sample you draw be representative?

The word "population" has a much broader meaning in statistics than in ordinary speech. Sometimes the word refers to a group of distinguishable individuals, such as all sixth-graders in Southern California, or all registered voters in the British Isles. In other instances it may refer to a set of repetitions—every time I flip this coin, or every time a leukemia patient is treated with cyclohexyl-chloroethyl-nitrosurea.

It's all too easy to get the two meanings confused. A psychologist friend of mine succeeded in training a mouse to run a maze, and then in running that same mouse—her name was Mehitabel—through some 100 planned variations of the basic course. "A sample of size 100, that's a lot. A lot of work, too," he confided.

Was it a large sample? He had run 100 variations of an experiment with a mouse named Mehitabel. And had obviously developed an extensive theory about Mehitabel's maze-running behavior. But as far as the general population of mice went, "You have a sample of size one," I told him.

Fortunately, he was able to use the concepts advanced in Chapter 8 of this book to design a more efficient experiment in which 10 different mice were each run through the same 10 mazes.

He did have to consult me again before he came up with the final design. The first time around, his idea was to repeat the initial experiment using Mehitabel's nine littermates. Do you see what's restrictive about this idea?

The same restriction would arise if one did a survey of political preferences by going from house to house interviewing the occupants while they sat around the dinner table. Their answers would not be the same as those that would be expressed in the privacy of the polling booth. People care what others think and when asked for an opinion about an emotionally charged topic in public, may or may not tell the truth. In fact, they're very unlikely to tell the truth if they think third parties are listening in. If a wife hears a husband say "Democrat," she may just reply "Democrat" herself in order to avoid an argument. To obtain a true representative sample, our observations must be independent of one another.

---

**Representative Samples**

Observations are independent.
Each item in the population has equal probability of being selected.
Identical method of measurement used on every item.
Observations are made in random order.

---

### 2.7.1  A Young Woman Tasting Herbal Tea

A young woman burst into my office last Tuesday evening, nostril flaring. "I can identify the correct brand of herbal tea 5 times out of 5."

"You're on," I said, "And I'll buy if you succeed."[9]

We went at once to the Student Union, where I proceeded to turn the whole coffee shop upside down searching for five matching cups. "What difference does the cup make?" I was asked. The difference is that I wanted this to be an experiment about herbal tea, not one about cups. I also made sure I had enough hot water for all five cups, and that all five brands were caffeine free, not mixtures of tea and herbs. Finally, I asked a second student if he wouldn't mind administering the test while I waited in the next room.

"Whatever for?" The answer to this question is that I didn't want my wizard of the tea bags to be able to discern from the expression on my face whether she was on the right track or not. I would prepare the tea in one room, remove the tea bags, then hand the cups to the second student. He would carry the cups to her and record her answers. Oh, and I had him stick a blindfold on her while I was in the next room preparing the tea.

---

[9]Friends (?) say it's extremely unlikely I ever made even this qualified offer to buy a round.

Do you feel this is a lot of effort to ensure the experiment measured exactly what she said it would? Not at all: I would have had to pay for all five cups if she'd got it right.[10]

---

**Warning**

Random is not the same as fair or equitable. Anything can happen and often does when samples are selected at random. Red appears 10 times in 10 consecutive spins of the roulette wheel. Impossible? Not at all. A result like this (or one equally improbable, like getting black ten times in a row) would be expected once in every 500 times.

---

## 2.7.2  Random Sampling and Representative Samples

An individual sample may not always be representative, that is, an exact copy of the population. But the method by which you sample must be representative in the sense it ensures that every possible sample is equally likely.

Random sampling would seem absolutely essential in the selection of jurors. The California Trial Jury Selection and Management Act[11] states, "It is the policy of the State of California that all persons selected for jury service shall be selected at random from the population of the area served by the court; that all qualified persons have an equal opportunity ... to be considered for jury service in the state and an obligation to serve as jurors when summoned for that purpose; and that it is the responsibility of jury commissioners to manage all jury systems in an efficient, equitable, and cost-effective manner in accordance with this chapter."

Can't we bend the rules a little? Do away with that blindfold or let the person recording the observations know exactly which cup of tea is which? The courts won't allow it and neither should we.

## 2.8  Summary

In this chapter, you learned to test a simple hypothesis by contrasting the observations as originally observed with their possible relabelings. You learned to frame both hypothesis and alternative and to accept the inevitability of error. The laws of probability were reviewed and applied to binomial observations and to collections of mutually exclusive events. You studied the concept of independence and learned the importance of drawing samples representative of the underlying population.

---

[10]She got only one out of five right, about what you'd expect by chance alone.

[11]Title 3, C.C.P. Section 191

---

### Do I Need to Take A Sample?

A friend of our family told me a moving story recently about the days before her late husband's death. "Our health insurance wouldn't cover the costs of his treatment for the final four years because the drug he was using wasn't approved."

"Four years is a long time," I replied, thinking with my head and not my heart, "you'd think the drug would have been approved after the first year or so, once they'd accumulated enough evidence with other patients."

She shook her head. "His drug never was approved; other patients had trouble with it apparently, but it helped my husband a lot."

I thought about the many things I might have said in rebuttal, about spontaneous remissions and the need for large controlled samples in which we examine the experiences of multiple patients under a variety of conditions, but I let it go. Words would not have eased her pain.

Her husband took a drug. He felt better for a period of four years. The two events might have been related or they might have been completely independent. Probably the latter, as the experiences of the vast majority of patients with this same drug had been negative.

The limited experience of one or two individuals can never constitute proof. Only when we examine a large representative sample can we begin to draw conclusions that extend to an entire population.

---

## 2.9  To Learn More

We'll learn a great deal more about framing and testing hypotheses in the next chapter. Stockmarr [2000] describes the framing of hypotheses in a courtroom setting. The tale of a lady tasting tea first appeared in Fisher [1935] along with some excellent advice on the design of experiments. The tale is revisited by Neyman [1950; pp. 272–294]. Ignore their advice at your peril; poor experimental design may lead to the so-called "spiral after-effect" and endless misinterpretation according to Stilson [1966]. Insight into the relationship between sample and population can be obtained from the opening chapters of Fisher [1925], Kempthorne [1955], and Hill [1966]. Hardy and Youse [1984] provide a careful review of probability fundamentals along with many exercises.

## 2.10  Exercises

**1.** Would you accept or reject the hypothesis that vitamin E increases longevity by three generations for the data of Section 2.1?

**2.** Suppose the police have DNA profiles for both the perpetrator of a crime and a recently arrested suspect.

**a.** What hypothesis would the prosecutor be concerned with?

**b.** Is this a null hypothesis?

**c.** What alternative would the defense be concerned with?

3. A majority of the doctors in North Carolina, a tobacco-growing state, find nothing wrong with smoking (or at least they didn't back in 1964). What hypotheses does this fact suggest? How would you go about testing them?

4. The Nielsen organization conducts weekly surveys of television viewing in the United States and issue estimates of the viewing audience for hundreds of T.V. programs.

**a.** What is the population from which their samples are drawn?

**b.** What would be an appropriate sampling method(s)?

**c.** What kinds of decisions are made based on these studies?

5. Suppose you have a six-sided die. What is the probability of rolling a 6? What is the probability of rolling a 6 twice in a row? What is the probability of rolling a pair of dice so the sum of their spots is 7? (Hint: add up the probabilities of the separate mutually exclusive outcomes.)

6. Are you psychic? Place a partition between you and a friend, so that you cannot see through the partition. Have your friend roll a die three times. Each time your friend rolls the die, you write down what you think was the result. Compare your guesses with what was actually observed. Would you be impressed if you got one out of three correctly? two out of three?

7. How many different subsamples of size 5 can be drawn without replacement from a sample of size 9? from a sample of size 10?

8. In how many different ways can one divide a sample of size 12 into four distinguishable subsamples?

9. In Section 8.4 we introduce the Latin Square in which $k$ treatments are laid our in $k$ rows and $k$ columns in such a way that no treatment appears twice in the same row and column. An example of a $4 \times 4 \times 4$ Latin Square would be

$$
\begin{array}{cccc}
A & B & C & D \\
D & A & B & C \\
C & D & A & B \\
B & C & D & A
\end{array}
$$

How many distinct $4 \times 4 \times 4$ Latin Squares are there?

10. An endocrinologist writing in the November 1973 issue of the *Ladies Home Journal* reported "I take strong exception with the prevailing medical opin-

ion that most 'tired' women are neurotic, bored, or sex-starved .... In my opinion, at least half such women can be shown to have a definite physical problem that causes their chronic illness." Are you convinced? Why or why not?

11. A study of 300 households in Santa Monica showed that a household produced an average of 4.5 pounds of garbage each day (0.8 pounds recyclable bottles, 1.0 pound yard clippings, 0.7 pounds paper, 2 pounds of yucky stuff).
    **a.** What is the sample in this study?
    **b.** How many observations in the sample?
    **c.** What are the variables?
    **d.** Do the results of this study sound as if they would be typical of your household?
    **e.** What is the population from which this sample is drawn?
    **f.** Would you be willing to extend inferences from this sample to this population?

12. Construct a theory to explain the opinion data recorded in Exercise 3 of Chapter 1. How would you go about selecting a sample to test your theory? Would you ask the same or different questions as those in that exercise?

13. "The Democrats will win this year." someone says. "How do you know?" you ask. "I interviewed 50 people and 29 favored the Democratic candidate." List all possible criticisms of this conclusion.

14. Draw the cumulative distributions of two sets of observations where the median of one set of observations is larger than the other.

15. Justice Blackmun notes in Ballew v. Georgia (1978) that if a minority group comprises 10% or less of a population, a jury of 12 persons selected at random from that population will fail to contain members of that minority at least 28% of the time. Show he is correct.

16. Do the following constitute independent observations?
    **a.** Number of abnormalities in each of several tissue sections taken from the same individual.
    **b.** Sales figures at Eaton's department store for its lamp and cosmetic departments.
    **c.** Sales figures at Eaton's department store for the months of May through November.
    **d.** Sales figures for the month of August at Eaton's department store and at its chief competitor Simpson-Sears.
    **e.** Opinions of several individuals whose names you obtained by sticking a pin through a phone book, and calling the "pinned" name on each page.

    **f.** Dow Jones Index and GNP of the United States.

    **g.** Today's price in Australian dollars of the German mark and the Japanese yen.

17. What is the probability that a family of three children has
    **a.** three boys given that it has at least one boy?
    **b.** three boys given that the last child was a boy? (Careful, children are not coin tosses.)

18. A card is drawn at random from a deck of 52 cards. What is the probability:
    **a.** It is a spade given it is black?
    **b.** It is higher than a ten given it is a club?
    **c.** It is a seven given it is lower than an eight?
    **d.** It is an odd-numbered card given it is black.

# CHAPTER  3

---

# Hypothesis Testing

In Chapter 2, we learned how to frame a hypothesis and an alternative and to use relabeling to generate a permutation distribution with which to discriminate between the two. In this chapter, you generate the permutation distributions for a variety of one- and two-sample hypotheses.

## 3.1   Five Steps to a Permutation Test

In this and succeeding chapters, you will apply permutation techniques to a wide variety of testing problems ranging from the simple to the complex. In each case, you will follow the same five-step procedure we followed in Section 2.2.

1. Analyze the problem—identify the hypothesis and the alternative(s) of interest.

2. Choose a test statistic.

3. Compute the test statistic for the original labeling of the observations.

4. Rearrange the labels, then recompute the test statistic. Repeat until you obtain the distribution of the test statistic for all possible rearrangements.

5. Accept or reject the hypothesis using this *permutation distribution* as a guide.

### 3.1.1   Analyze the Problem

Recall I wanted to assess the life-extending properties of a new experimental treatment with vitamin E. My null hypothesis was that the growth potential of a culture of human diploid fibroblasts would not be affected by the presence of vitamin E in the media. All the cultures would have equal growth potential. The alternative of interest was that cells grown in the presence of vitamin E would be capable of many more cell divisions.

## 3.1.2  Choose a Test Statistic

The next step in the permutation method is to choose a test statistic that discriminates between the hypothesis and the alternative. In Chapter 2, I chose as my test statistic the sum of the observations in the group treated with vitamin E. If the alternative is true, and vitamin E has a beneficial effect, this sum would be larger most of the time than the sum of the counts in the untreated group. If the null hypothesis is true, that is, if it doesn't make any difference which treatment cells receive, then the sums of the two groups of observations would be approximately the same. One sum might be smaller or larger than the other by chance, but in the vast majority of cases the two shouldn't be all that different.

## 3.1.3  Compute the Test Statistic

The first three observations my technician showed me—121, 118, and 110—belonged to the cultures that received vitamin E. Their sum is 349.

## 3.1.4  Rearrange the Observations

Reproduced below from Section 2.2 are the 20 different possible reassignments of the six labels to the six observations along with the values of the test statistic associated with each reassignment.

|  | First Group | | | Second Group | | | Sum of First Group |
|---|---|---|---|---|---|---|---|
| 1. | 121 | 118 | 110 | 34 | 22 | 12 | 349 |
| 2. | 121 | 118 | 34 | 110 | 22 | 12 | 273 |
| 3. | 121 | 110 | 34 | 118 | 22 | 12 | 265 |
| 4. | 118 | 110 | 34 | 121 | 22 | 12 | 262 |
| 5. | 121 | 118 | 22 | 110 | 34 | 12 | 261 |
| 6. | 121 | 110 | 22 | 118 | 34 | 12 | 253 |
| 7. | 121 | 118 | 12 | 110 | 34 | 22 | 251 |
| 8. | 118 | 110 | 22 | 121 | 34 | 12 | 250 |
| 9. | 121 | 110 | 12 | 118 | 34 | 22 | 243 |
| 10. | 118 | 110 | 12 | 121 | 34 | 22 | 240 |
| 11. | 121 | 34 | 22 | 118 | 110 | 12 | 177 |
| 12. | 118 | 34 | 22 | 121 | 110 | 12 | 174 |
| 13. | 121 | 34 | 12 | 118 | 110 | 22 | 167 |
| 14. | 110 | 34 | 22 | 121 | 118 | 12 | 166 |
| 15. | 118 | 34 | 12 | 121 | 110 | 22 | 164 |
| 16. | 110 | 34 | 12 | 121 | 118 | 22 | 156 |
| 17. | 121 | 22 | 12 | 118 | 110 | 34 | 155 |
| 18. | 118 | 22 | 12 | 121 | 110 | 34 | 152 |
| 19. | 110 | 22 | 12 | 121 | 118 | 34 | 144 |
| 20. | 34 | 22 | 12 | 121 | 118 | 110 | 68 |

## 3.1.5   Draw a Conclusion

Again, we see that the sum of the observations in the original vitamin-E-treated group, 349, is equaled only once and never exceeded in the 20 distinct relabelings. If chance alone is operating, then such an extreme value is a rare, only-1-time-in-20 event. We can reject the null hypothesis and embrace the alternative that the treatment is effective and responsible for the observed difference, knowing that we risk making an error only once in 20 times. Such a test is said to be at the 5% (1 in 20) *significance level.*

---

**Five Steps to a Permutation Test**

1. Analyze the problem:

    (a) What is the hypothesis? What are the alternatives?
    (b) How are the data distributed?
    (c) What losses are associated with bad decisions?

2. Choose a statistic that will distinguish the hypothesis from the alternative.

3. Compute the test statistic for the original observations.

4. Derive the permutation distribution of the test statistic:

    (a) Rearrange (relabel) the observations.
    (b) Compute the test statistic for the new arrangement.
    (c) Compare the new value of the test statistic with the value obtained for the original observations.
    (d) Repeat steps a, b, and c until you are ready to make a decision.

5. Draw a conclusion:

    Reject the hypothesis and accept the alternative if the value of the test statistic for the observations as they were labeled originally is an extreme value in the permutation distribution of the statistic.

    Accept the hypothesis and reject the alternative, otherwise.

---

## 3.2   A Second Example

Suppose once again we have two samples: The first, control sample takes values 0, 1, 2, 3, 19. The second, treatment sample takes values 3.1, 3.5, 4, 5, and 6. Does the treatment have an effect?

In technical terms, we want to test that the median of the control population (or the mean, or the population distribution itself) is the same as the median (mean, distribution) of the vitamin-E-treated population, against the alternative that the median (mean) is higher or the frequency distribution is shifted to the right toward higher values.[1]

The answer would be immediate if it were not for the 19 in the first sample. The presence of this extreme value increases the mean of the first sample from 1.5 to 5. What if this "19" is really a "1.9"? Should a simple typographical error be allowed to affect our conclusions?

To dilute the effect of this extreme value on the results, we convert all the data to ranks, giving the smallest observation a rank of 1, the next smallest the rank of 2, and so forth. The first sample includes the ranks 1, 2, 3, 4, 10 and the second sample includes the ranks 5, 6, 7, 8, 9. Is the second sample drawn from a different population than the first?

Let's count. The sum of the ranks in the first sample is 20. All the rearrangements with first samples of the form 1, 2, 3, 4, k, where k is chosen from the values {5, 6, 7, 8, 9, or 10} have sums that are as small or smaller than that of our original sample. That's a total of six rearrangements. The four rearrangements whose first sample contains 1, 2, 3, 5, and a fifth number chosen from the set {6, 7, 8, 9} also have sums that are as small or smaller. That's 6 + 4 = 10 rearrangements so far. Continuing in this fashion—I leave the complete enumeration for you as an exercise—we find 13 of the $\binom{10}{5} = 252$ possible rearrangements have sums that are as small or smaller than the sum of the observations in our original sample. Two samples this different will be drawn from the same population just over 5% of the time by chance.

| Group 1 | | | | | Group 2 | | | | | Sum of Group 1 |
|---|---|---|---|---|---|---|---|---|---|---|
| 1 | 2 | 3 | 4 | 10 | 5 | 6 | 7 | 8 | 9 | 20 |
| 1 | 2 | 3 | 4 | 5 | 6 | 7 | 8 | 9 | 10 | 15 |
| 1 | 2 | 3 | 4 | 6 | 5 | 7 | 8 | 9 | 10 | 16 |
| 1 | 2 | 3 | 4 | 7 | 5 | 6 | 8 | 9 | 10 | 17 |
| 1 | 2 | 3 | 4 | 8 | 5 | 6 | 7 | 9 | 10 | 18 |
| 1 | 2 | 3 | 4 | 9 | 5 | 6 | 7 | 8 | 10 | 19 |
| 1 | 2 | 3 | 5 | 6 | 4 | 6 | 7 | 8 | 10 | 17 |
| 1 | 2 | 3 | 5 | 7 | 4 | 6 | 8 | 9 | 10 | 18 |
| and so forth | | | | | | | | | | |

## 3.2.1  Missing Data

Suppose we've designed an experiment with four individuals in each of two treatment groups, but when the results are in, we learn that one of the individuals in the second treatment group failed to show for his appointment, so we only have seven observations in all. Do we examine the 7 choose 3 rearrangements ignoring the missing observation all together? Or do we examine all 8 choose 4 rearrangements computing the mean of the first sample each time? Our answer depends

---

[1] See Exercise 14.

on whether the observation is missing at random or whether it is more likely to be missing in one treatment group rather than another. If it is missing at random, then we should examine all 8 choose 4 rearrangements. If the proportion of missing values depends on the treatment, no statistical analysis is justifiable until we have looked more deeply into the underlying causes.

### 3.2.2   More General Hypotheses

Our permutation technique is applicable whenever we can freely exchange labels; this is obviously the case when testing the hypothesis that both samples are drawn from the same population. But suppose our hypothesis is somewhat different, that we believe, for example, that a new gasoline additive will increase mileage by at least 20 miles per tank. (Otherwise, why pay the big bucks for the additive?) Before using the additive, we recorded 295, 320, 329, and 315 miles per tank of gasoline under various traffic conditions. With the additive, we recorded 330, 310, 345, and 340 miles per tank.

A permutation test will still be applicable providing the only difference between the before and after distributions from which these samples are drawn is a shift in means. If some other aspect of the mileage distribution has changed as a result of the new additive, for example, if our gas mileage is no longer as variable, then we should not perform a permutation test (see Section 3.2.3). To perform the analysis, we begin by subtracting 20 from each of the post-additive figures. Our revised hypothesis is that the transformed mileages will be approximately the same, versus the alternative that the post-additive transformed mileages will be less than the others. Our two transformed samples are

$$
\begin{array}{cccc}
295 & 320 & 329 & 315 \\
305 & 290 & 325 & 320
\end{array}
$$

The actual analysis is left to you as an exercise.

### 3.2.3   Behrens-Fisher Problem

A slight modification of the preceding assumptions yields a problem whose exact solution has eluded statisticians for more than half a century. Suppose we cannot be certain that the variances of the two populations from which the samples are drawn are the same. Can we still test for the equality of the means? We cannot use the permutation approach because the labels are no longer exchangeable under the null hypothesis. For even if the null hypothesis were true, an observation in the first sample could be distinguished from one in the second sample because it comes from a population with a different variance. We can use the bootstrap described in Section 1.6.3; after establishing some preliminary definitions, we return in Section 5.2.5 to the solution of what is known as the Behrens-Fisher problem.

---

**Two-Sample Test for Location**

Hypothesis H: mean/medians of groups differ by $d$.
Alternative K: mean/medians of groups differ by $\delta + d$, where $\delta > 0$.
Assumptions:

1. Under the hypothesis, the labels on the observations can be exchanged.

2. All observations in the first sample come from the same distribution $F$.

3. All observations in the second sample come from the distribution $G$ where $G[x] = \Pr\{Y_k \leq x\} = \Pr\{X_j \leq x - \delta\} = F[x - \delta]$ for all $x$.

Procedure:
Transform the observations in the first sample by subtracting $\delta$ from each.
Compute the sum of the observations in the smallest sample.
Compute this sum for all possible rearrangements of the combined sample to form the permutation distribution of the test statistic.

Draw a conclusion:
Reject the hypothesis in favor of the alternative if the original sum is an extreme value of the permutation distribution.

---

## 3.3   Comparing Variances

Precision is essential in a manufacturing process. Items that are too far out of tolerance must be discarded and an entire production line brought to a halt if too many items exceed (or fall below) designated specifications. With some testing equipment, such as that used in hospitals, precision can be more important than accuracy. Accuracy can always be achieved through the use of standards with known values, while a lack of precision may render an entire sequence of tests invalid.

There is no shortage of methods to test the hypothesis that two samples come from populations with the same inherent variability, but few can be relied on. Many methods promise an error rate (significance level) of 5%, but in reality make errors as frequently as 8% to 20% of the time. Other methods for comparing variances have severe restrictions.

For example, a permutation test based on the ratio of the sample variances is appropriate only if the means of the two populations are the same or we know their values. If the means are the same, then the labels are exchangeable under the null hypothesis. And if we know the value of the means, then we can make a preliminary transformation that makes the labels exchangeable.

We can derive a permutation test for comparing variances that is free of these restrictions if, instead of working with the original observations, we replace them with the differences between successive order statistics and then permute the la-

---

### When Are Labels Exchangeable?

To perform a permutation test, the labels on the observations must be exchangeable under the null hypothesis. They are exchangeable whenever the joint distribution of the observations is unaffected by exchanges of the labels, for example, when

$$\Pr\{x \leq a \text{ and } y \leq b \text{ and } z \leq c, \ldots\} = \Pr\{y \leq a \text{ and } x \leq b \text{ and } z \leq c, \ldots\}$$
$$= \Pr\{z \leq a \text{ and } x \leq b \text{ and } y \leq c, \ldots\}$$

Thus, they are exchangeable when all variables come from the same distribution and are independent.

---

bels. The test statistic proposed by Aly [1990] is

$$\delta = \sum_{i=1}^{m-1} i(m-i)(X_{(i+1)} - X_{(i)}),$$

where $X_{(1)} \leq X_{(2)} \leq \ldots \leq X_{(m)}$ are the order statistics of the first sample. That is, $X_{(1)}$ is the smallest of the observations in the first sample (the minimum), $X_{(2)}$ is the second smallest, and so forth, up to $X_{(m)}$ the maximum.

To illustrate the application of Aly's statistic, suppose the first sample consists of the measurements 121, 123, 126, 128.5, 129 and the second sample of the measurements 153, 154, 155, 156, 158. $X_{(1)}=121$, $X_{(2)}=123$, and so forth.

Set $z_{1i} = X_{(i+1)} - X_{(i)}$ for $i = 1, \ldots, 4$. In this instance, $z_{11} = 123 - 121 = 2$, $z_{12} = 3$, $z_{13} = 2.5$, $z_{14} = 0.5$.

The original value of Aly's test statistic is $8 + 18 + 15 + 2 = 38$. To compute the test statistic for other arrangements, we also need to know the differences $z_{2i} = Y_{(i+1)} - Y_{(i)}$ for the second sample; $z_{21} = 1$, $z_{22} = 1$, $z_{23} = 1$, $z_{24} = 2$.

Only certain exchanges are possible. Rearrangements are formed by first choosing either $z_{11}$ or $z_{21}$, next either $z_{12}$ or $z_{22}$, and so forth until we have a set of four differences.

One possible rearrangement is $\{2, 1, 1, 2\}$, which yields a value of $\delta = 20$. There are $2^4 = 16$ rearrangements in all, of which only one $\{2, 3, 2.5, 2\}$ yields a more extreme value of the test statistic than our original observations. With two out of 16 rearrangements yielding values of the statistic as or more extreme than the original, we should accept the null hypothesis. Better still, given the limited number of possible rearrangements, we should gather more data before we make a decision.[2]

---

[2]How much data? See Chapter 6.

**Monte Carlo Sampling**

Determining the significance of a set of observations by looking at all possible relabelings can be quite time-consuming for even medium-sized samples. Two samples of size three have only 20 possible rearrangements. But two samples of size six have 924, and two samples of size 10 have 184,756. We could program our computer to run all night long (see Appendix 1) but there is a better way.

With a Monte Carlo, we restrict our attention to a random sample of rearrangements. I usually begin with about 400. Let $p*$ denote the resultant estimate of the true significance level $p$. $p*$ has a binomial distribution (see Section 2.5.2) with probability $p$ and 400 trials. If $p$ is 5% then 95% of the time we would expect to have values of $p*$ within about half a percent of 5%, that is, in the interval (4.5%,5.5%).

If a more precise estimate is required, then we need to look at additional random rearrangements. With 1600 random rearrangements, then 95% of the time when $p = 5\%$ we would expect to find values of $p*$ in the interval (4.75%, 5.25%).

### 3.3.1    Unequal Sample Sizes

If our second sample is larger than the first, we have to resample in two stages. Suppose $m$ observations are in the first sample and $n$ in the second, where $m < n$. Select a random subset of $m$ values $\{Y_{*i}, i = 1, \ldots, m\}$ without replacement from the $n$ observations in the second sample. Compute the order statistics $Y_{*(1)} \leq Y_{*(2)} \leq \ldots$, their differences $\{z_{*2i}\}$, and the value of Aly's measure of dispersion for the $2^m$ possible rearrangements of the combined sample $\{\{z_{1i}\}, \{z_{*2i}\}\}$. Repeat this procedure for all possible subsets of the second sample, combine all the permutation distributions into one, and compare Aly's measure for the original observations with the combined distribution.

Should the total number of calculations appear prohibitive were all subsets used, then use only a representative random sample of them.

## 3.4    Pitman Correlation

We can easily generalize from two samples to three or more if the samples are ordered in some way, by the dose administered, for example.

Frank, Trzos, and Good [1977] studied the increase in chromosome abnormalities and micronuclei as the dose of various compounds known to cause mutations was increased. Their object was to develop an inexpensive but sensitive biochemical test for mutagenicity that would be able to detect even marginal effects. The results of their experiment are reproduced in Table 3.1.

To analyze such data, Pitman [1937] proposes a test for linear correlation with three or more ordered samples using as test statistic $S = \Sigma g[i]x_i$ where $x_i$ is the

Table 3.1. Micronuclei in Polychromatophilic Erythrocytes and Chromosome Alterations in the Bone Marrow of Mice Treated With CY

| Dose (mg/kg) | Number of Animals | Micronuceli per 200 cells | | | | | Breaks per 25 cells | | | | |
|---|---|---|---|---|---|---|---|---|---|---|---|
| 0  | 4 | 0 | 0 | 0 | 0 |    | 0 | 1 | 1 | 2 |   |
| 5  | 5 | 1 | 1 | 1 | 4 | 5  | 0 | 1 | 2 | 3 | 5 |
| 20 | 4 | 0 | 0 | 0 | 4 |    | 3 | 5 | 7 | 7 |   |
| 80 | 5 | 2 | 3 | 5 | 11 | 20 | 6 | 7 | 8 | 9 | 9 |

sum of the observations in the $i$th dose group, and $g[i]$ is any monotone increasing function.[3] The simplest example of such a function is $g[i] = i$, with test statistic $S = \Sigma g[i]x_i$. In this instance, we take $g[\text{dose}] = \log[\text{dose} + 1]$, as the anticipated effect is proportional to the logarithm of the dose.[4] Our test statistic is $S = \Sigma \log[\text{dose}_i + 1]x_i$.

The original data for breaks may be written in the form

$$0\ 1\ 1\ 2 \quad 0\ 1\ 2\ 3\ 5 \quad 3\ 5\ 7\ 7 \quad 6\ 7\ 8\ 9\ 9$$

As $\log[0 + 1] = 0$, the value of the Pitman statistic for the original data is $0 + 11 * \log[6] + 22 * \log[21] + 39 * \log[81] = 112.1$. The only larger values are associated with the small handful of rearrangements of the form

$$
\begin{array}{llll}
0\ 0\ 1\ 2 & 1\ 1\ 2\ 3\ 5 & 3\ 5\ 7\ 7 & 6\ 7\ 8\ 9\ 9 \\
0\ 0\ 1\ 1 & 1\ 2\ 2\ 3\ 5 & 3\ 5\ 7\ 7 & 6\ 7\ 8\ 9\ 9 \\
0\ 0\ 1\ 1 & 1\ 2\ 2\ 3\ 3 & 5\ 5\ 7\ 7 & 6\ 7\ 8\ 9\ 9 \\
0\ 0\ 1\ 2 & 1\ 1\ 2\ 3\ 3 & 5\ 5\ 7\ 7 & 6\ 7\ 8\ 9\ 9 \\
0\ 1\ 1\ 2 & 0\ 1\ 2\ 3\ 3 & 5\ 5\ 7\ 7 & 6\ 7\ 8\ 9\ 9 \\
0\ 1\ 1\ 2 & 0\ 1\ 2\ 3\ 5 & 3\ 5\ 6\ 7 & 7\ 7\ 8\ 9\ 9 \\
0\ 0\ 1\ 2 & 1\ 1\ 2\ 3\ 5 & 3\ 5\ 6\ 7 & 7\ 7\ 8\ 9\ 9 \\
0\ 0\ 1\ 1 & 1\ 2\ 2\ 3\ 5 & 3\ 5\ 6\ 7 & 7\ 7\ 8\ 9\ 9 \\
0\ 0\ 1\ 1 & 1\ 2\ 2\ 3\ 3 & 5\ 5\ 6\ 7 & 7\ 7\ 8\ 9\ 9 \\
0\ 0\ 1\ 2 & 1\ 1\ 2\ 3\ 3 & 5\ 5\ 6\ 7 & 7\ 7\ 8\ 9\ 9 \\
0\ 1\ 1\ 2 & 0\ 1\ 2\ 3\ 3 & 5\ 5\ 6\ 7 & 7\ 7\ 8\ 9\ 9 \\
\end{array}
$$

As there are $\binom{18}{454}$ rearrangements in all,[5] a statistically significant ordered dose response ($p < 0.001$) has been detected. The micronuclei also exhibit a statistically significantly dose response when we calculate the permutation distribution of $S = \Sigma \log[\text{dose}_i + 1]n_i$. To make the calculations, we took advantage of the C++ computer program detailed in Appendix 1; the only change was in the subroutine used to compute the test statistic.

---

[3] A monotone increasing function $g[x]$ increases steadily as $x$ goes larger. The simplest example is $g[i] = i$. Another is $g[i] = ai + b$, with $a > 0$.

[4] Adding a 1 to the dose keeps this function from blowing up at a dose of zero.

[5] This is the number of ways that 18 things can be divided among four subgroups of sizes 4, 5, 4, and 5. It is equal to 18!/4!5!4!5!

A word of caution: if we use some function of the dose other than $g[\text{dose}] = \log[\text{dose} + 1]$, we might observe a different result. Our choice of a test statistic must always make practical as well as statistical sense.

---

**k-Sample Test for Ordered Samples**

Hypothesis H: all distributions and all population means are the same.
  Alternative K: the population means are ordered.
Assumptions:

1. Labels on the observations can be exchanged if the hypothesis is true.

2. All the observations in the $i^{th}$ sample come from the same distribution $G_i$, where $G_i[x] = \Pr\{X_{ij} \leq x\} = F[x - \delta]$.

Test statistic:
  $S = \Sigma g[i] x_{i.}$ where $x_{i.}$ is the sum of the observations in the $i$th sample.

---

## 3.4.1  Effect of Ties

Ties can complicate the determination of the significance level. Because of ties, each of the rearrangements noted in the preceding example might actually have resulted from several distinct reassignments of subjects to treatment groups and must be weighted accordingly. To illustrate this point, suppose we put tags on the 1's in the original sample

$$0\,1^*\,1\#\,2 \quad 0\,1\,2\,3\,5 \quad 3\,5\,7\,7 \quad 6\,7\,8\,9\,9$$

The rearrangement

$$0\,0\,1\,2 \quad 1\,1\,2\,3\,5 \quad 3\,5\,7\,7 \quad 6\,7\,8\,9\,9$$

corresponds to the three reassignments

$$0\,0\,1\,2 \quad 1^*\,1\#\,2\,3\,5 \quad 3\,5\,7\,7 \quad 6\,7\,8\,9\,9$$
$$0\,0\,1^*\,2 \quad 1\,1\#\,2\,3\,5 \quad 3\,5\,7\,7 \quad 6\,7\,8\,9\,9$$
$$0\,0\,1\#\,2 \quad 1\,1^*\,2\,3\,5 \quad 3\,5\,7\,7 \quad 6\,7\,8\,9\,9$$

The 18 observations are divided into four dose groups containing 4, 5, 4, and 5 observations, respectively, so that there are $\binom{18}{454}$ possible reassignments of observations to dose groups. Each reassignment has probability $1/\binom{18}{454}$ of occurring, so the probability of the rearrangement

$$0\,0\,1\,2 \quad 1\,1\,2\,3\,5 \quad 3\,5\,7\,7 \quad 6\,7\,8\,9\,9$$

is $3/\binom{18}{454}$.

To determine the significance level when there are ties, weight each distinct rearrangement by its probability of occurrence. This weighting is done automatically if you use Monte Carlo sampling methods.

## 3.5   Bivariate Dependence

Are two variables correlated with one another? Is there a causal relationship be-tween them or between them and a third hidden variable? Take another look at Figure 1.4 in Chapter 1. Surely arm span and height must be closely related. We can use the Pitman Correlation method introduced in the previous section to ver-ify our conjecture. If $\{(a_1, h_1), (a_2, h_2), \ldots (a_n, h_n)\}$ is a set of $n$ observations on arm span and height, then to test for association between them we need to look at the permutation distribution of the Pitman Correlation

$$S = \sum_{i=1}^{n} a_i h_i.$$

To illustrate this approach, let's look at the arm span and height of the five shortest students in my sixth grade class

$$(139, 137) \quad (140, 138.5) \quad (141, 140) \quad (142.5, 141) \quad (143.5, 142)$$

Both the arm spans and the heights are in increasing order. Is this just coinci-dence? What is the probability that an event like this could happen by chance alone? We could list all possible permutations of both arm span and height, but this won't be necessary. We can get exactly the same result if we fix the order of one of the variables, the height, for example, and look at the 5! = 120 ways in which we could rearrange the arm span readings:

$$(140, 137) \quad (139, 138.5) \quad (141, 140) \quad (142.5, 141) \quad (143.5, 142)$$
$$(141, 137) \quad (140, 138.5) \quad (139, 140) \quad (142.5, 141) \quad (143.5, 142)$$

and so forth.

Obviously, the arrangement we started with is the most extreme, occurring ex-actly one time in 120 by chance alone. Applying this same test to all 22 pairs of observations, we find the odds are less than 1 in a million that what we observed occurred by chance alone and conclude that arm span and height are directly re-lated.

## 3.6   One-Sample Tests

The resampling methods that work for two, three, or more samples, won't work for one. Another approach is required. In this section, we consider two al-ternate approaches to the problem of testing a hypothesis concerning a single population—the bootstrap and the permutation test.

### 3.6.1   The Bootstrap

Suppose we wish to test the hypothesis that the mean temperature of a process is 440 °C and have already taken the 20 readings shown in Table 3.2. To apply

---

**Test for Bivariate Dependence**

H: The observations within each pair are independent of one another.
K: The observations within each pair are all positively correlated or are all negatively correlated.

Assumptions:
1. $(x_i, y_i)$ denotes the members of the $i^{\text{th}}$ pair of observations.
2. The labels of the $\{x_i\}$ are exchangeable as are those of the $\{y_i\}$.

Test statistic:
$S = \Sigma y_i x_i$. Reject the hypothesis of independence if the value of $S$ for the observations before they are rearranged is an extreme value in the permutation distribution.

---

Table 3.2. Process Temperature (from Cox and Snell, 1981)

431, 450, 431, 453, 481, 449, 441, 476, 460, 482, 472, 465, 421, 452, 451, 430, 458, 446, 466, 476.

---

the bootstrap introduced in Section 1.6.4, we need only assume the 20 readings are independent. The mean reading is 454.6. This represents a deviation from our hypothesized mean temperature of 14.6 degrees. How likely is a deviation this large to arise by chance? To find out, we drew 100 bootstrap resamples. The results are depicted in Table 3.3 and Figure 3.1.

In not one of these 100 bootstrap samples, did the mean fall below 444.95. We reject the hypothesis that the mean of our process is 440 ° C.

One caveat. This test is not an exact one; the actual significance level may be greater than 1%. (Tips on reducing the level error will be found in Chapter 5.) This approach also can be used for testing whether the variance, the median, or some other attribute of the distribution takes on a specific value.

## 3.6.2   Permutation Test

We can achieve an exact significance level using the permutation approach if we are able to make two assumptions: i) the readings in Table 3.3 are independent; ii) they come from a symmetric distribution.

First, we create a table of the deviations about 440, the hypothesized process mean (Table 3.4). Right off the bat, we note something suspicious: If the underlying distribution is symmetric so that positive and negative deviations are equally

444.95                    Medians of Bootstrap Samples                    464.55

**Figure 3.1.**   Scatter plot of bootstrap median temperatures.

Table 3.3. Percentiles of the Resampling Distribution

| 1% | 445.15 | 5% | 447.8 | 10% | 449.42 |
|-----|--------|-----|--------|------|--------|
| 25% | 452.1 | 50% | 454.3 | 75% | 456.6 |
| 90% | 458.98 | 95% | 459.52 | 99% | 462.7 |

Table 3.4.  Process Temperature Deviations from a Hypothesized Mean Value of 440

| 431 | +9 | 476 | +36 | 451 | +11 |
|-----|-----|-----|-----|-----|-----|
| 450 | +10 | 460 | +20 | 430 | −10 |
| 431 | −9 | 482 | +42 | 458 | +18 |
| 453 | +13 | 472 | +32 | 446 | +6 |
| 481 | +41 | 465 | +25 | 466 | +26 |
| 449 | +9 | 421 | −19 | 476 | +36 |
| 441 | +1 | 452 | +12 | | |

likely how is it that only 4 of the 20 deviations are negative? This can happen less than one in a 100 times by chance (see Section 2.5), and we should reject our hypothesis that the process mean is 440.

Our test is a sign test based on the binomial, rather than a permutation test, but why waste time when you don't need to?

The sign test is the optimal approach when observations are dirt cheap. When data are expensive or difficult to obtain, we may need a more powerful test.

To more fully illustrate the permutation test for a single sample, let's consider a test of a hypothesized mean value of 450, again using the data of Table 3.2.

Now the decision isn't so obvious. Examining Table 3.5, we see that seven of the 20 deviations are negative, and there are large negative deviations as well as large positive ones.

Suppose we'd lost track of the signs (the way we lost track of the labels in my earlier example). If we can assume the underlying distribution is symmetric about zero—our fundamental assumption—so that positive and negative values are equally likely, we can attach new signs at random, selecting a plus or minus

Table 3.5. Process Temperature Deviations from a Hypothesized Mean Value of 450

| 431 | −19 | 476 | +26 | 451 | +1 |
|-----|-----|-----|-----|-----|-----|
| 450 | +0 | 460 | +10 | 430 | −20 |
| 431 | −19 | 482 | +32 | 458 | +8 |
| 453 | +3 | 472 | +22 | 446 | −4 |
| 481 | +31 | 465 | +15 | 466 | +16 |
| 449 | −1 | 421 | −29 | 476 | +26 |
| 441 | −9 | 452 | +2 | | |

with equal probability, for a total of $2 \times 1 \times 2 \times 2 \times \cdots$ or $2^{19}$ possible rearrangements. (The 0 doesn't get rearranged.) The sum of the negative differences for our current rearrangement is $-101$. There are far too many rearrangements, $2^{20}$ or one million million, for us to even attempt a count. A random sample of 10,000 rearrangements yields an estimate of the probability of a sum of negative differences as small as that observed in Table 3.5 on the order of 13%. We accept the hypothesis that the mean is 450.

## 3.7   Matched Pairs

We can immediately apply our results for a single sample to matched pairs. In a matched-pairs experiment, each subject in the treatment group is matched as closely as possible by a subject in the control group. For example, if a 45-year-old black male hypertensive is given a blood-pressure-lowering pill, then we give a second similarly built 45-year-old black male hypertensive a placebo.[6] One member of each pair is then assigned at random to the treatment group, and the other member is assigned to the controls.

The value of this approach is that it narrows our focus to differences arising solely from treatment, reducing, through matching, the "noise" or dispersion resulting from differences among individuals.

Assuming we've been successful in our matching, we end up with a series of independent pairs of observations $(X_i, Y_i)$ where the members of each pair have been drawn from distributions that have been shifted with respect to one another by a constant amount $\delta$, so that $\Pr\{Y_i \leq x\} = \Pr\{X_i \leq x - \delta\} = F[x - \delta]$. Regardless of the form of this unknown distribution F, the differences $Z_i = X_i - Y_i$ will be symmetrically distributed about the unknown shift parameter $\delta$.

### 3.7.1   An Example: Pre-Treatment and Post-Treatment Levels

In this example, pre-treatment and post-treatment serum antigen levels were measured in 20 AIDS patients. The data of Table 3.6 was entered into StatXact® and a permutation test for two related samples used to test whether there is a statistically significant effect due to treatment.

The result of the analysis with StatXact® $p = 0.0021$, represents a highly statistically significant difference.[7] If we use Student's $t$ in this example, as some textbooks recommend (see Section 4.3.2), we would obtain a one-sided $p$-value of 0.0413 and a two-sided $p$-value of 0.0826. Which test is correct? The significance levels derived from Student's $t$ are exact when the underlying observations are independent and each is drawn from the same normal distribution. The permutation test is exact even if the underlying distribution is not normal as long as the observations are independent and identically distributed.

---

[6] A placebo looks and tastes like a medicine but consists of an innocuous chemical or filler without any real biological effect. Does it work? Many patients get better with the help of a placebo, especially if a doctor, a witch doctor, a parent, or a significant other gives it to them.

[7] Which may or may not be of biological significance.

Table 3.6. Response of Serum Antigen Level (pg/ml) to AZT[#]

| Patient ID | Pre-treatment | Post-treatment | Difference |
|---|---|---|---|
| 01 | 149 | 0 | −149 |
| 02 | 0 | 51 | 51 |
| 03 | 0 | 0 | 0 |
| 04 | 259 | 385 | 126 |
| 05 | 106 | 0 | −106 |
| 06 | 255 | 235 | −20 |
| 07 | 0 | 0 | 0 |
| 08 | 52 | 0 | −52 |
| 09 | 340 | 48 | −292 |
| 10 | 0 | 65 | 65 |
| 11 | 180 | 77 | −103 |
| 12 | 0 | 0 | 0 |
| 13 | 84 | 0 | −84 |
| 14 | 89 | 0 | −89 |
| 15 | 212 | 53 | −159 |
| 16 | 554 | 150 | −404 |
| 17 | 500 | 0 | −500 |
| 18 | 424 | 165 | −259 |
| 19 | 112 | 98 | −14 |
| 20 | 2, 600 | 0 | −2, 600 |
| # Source: Makutch and Parks [1988] | | | |

**Permutation Test for Two Related Samples with StatXact**

Summary of Exact distribution of PERMUTATION TEST statistic:

| Min | Max | Mean | Std-dev | Observed | Standardized |
|---|---|---|---|---|---|
| 0.0000 | 5108. | 2554. | 1369. | 177.0 | −1.736 |

Exact Inference:
One-sided $p$-value:

$$\Pr\{\text{Test Statistic .LE. Observed}\} = 0.0011$$
$$\Pr\{\text{Test Statistic .EQ. Observed}\} = 0.0000$$

Two-sided $p$-value:

$$\Pr\{|\text{Test Statistic - Mean}| \text{ .GE. } |\text{Observed - Mean}|\} = 0.0021$$

Two-sided $p$-value: $2 *$ One-Sided $= 0.0021$

---

**Permutation Test for Matched Pairs**

Hypothesis H: Means/medians of the members of each pair are the same.
Alternative K: Means/medians of the members of each pair differ by $\delta > 0$.

Assumptions:

1. Labels within each pair of observations can be exchanged if the hypothesis is true; that is, under the hypothesis, pre- and post-treatment values are identically distributed.

2. Pairs of observations are independent of one another.

Test Statistic:
$$\sum\nolimits_{x_i > y_i} (x_i - y_i)$$

---

## 3.8   Summary

In this chapter you learned the five steps necessary to perform a permutation test along with specific tests for comparing the locations and dispersions of two populations. Pitman Correlation, which you studied here, enables you to compare the locations of several populations against an ordered alternative and to test for bivariate dependence. You used two different resampling methods—the permutation test and the bootstrap—to test hypotheses regarding the location parameter of a single population. And you saw you could use these same tests to compare two related populations via matched pairs.

## 3.9   To Learn More

The permutation tests were introduced by Pitman [1937, 1938] and Fisher [1935] and have seen wide application in anthropology (Valdes-Perez and Pericliev, 1999), chemistry (vanKeerberghen et al, 1991), climatology (Hisdal et al., 2001), clinical trials (Berger, 2000; Grossman et al., 2000; Gail, Tan, and Piantadosi, 1988; Howard, 1981), cybernetics (Valdesperez, 1995), ecology (Busby, 1990; Cade, 1997; Pollard, Lakhand and Rothrey, 1987; Prager and Hoenig, 1989), genetics (Thaler-Neto, Fries and Thaller, 2000; Gonser et al., 2000), law (Gastwirht, 1992), education (Gliddentracy and Greenwood, 1997; Gliddentracy and Parraga, 1996), medicine (Feinstein, 1973; Tsutakawa and Yang, 1974), meteorology (Gabriel, 1979; Tukey, 1985), neurology (Burgess and Gruzelier, 2000; Ford, Colom, and Bland, 1989), ornithology (Cade and Hoffman, 1993), pharmacology (Plackett and Hewlett, 1963), physiology (Boess et al., 1990; Faris and Sainsbury, 1990;

Zempo et al, 1996), psychology (Antretter, Dunkel, and Haring, 2000; Hollander and Penna, 1988; Hollander and Sethuraman, 1978; Kazdin, 1976,1980), radiology (Milano, Maggi and del Turco, 2000), sociology (Tsuji, 2000), software engineering (Laitenberger et al., 2000), theology (Witzum, Rips, and Rosenberg, 1994), toxicology (Farrar and Crump, 1988, 1991), virology (Good, 1979) and zoology (Adams and Anthony, 1996; Jackson, 1990). Texts dealing with their application include Bradley [1968], Edgington [1995], Maritz [1996], Noreen [1989], Manly [1997], and Good [2000]. See, also, the review by Barbella, Deby, and Glandwehr [1990]. Early articles include Pearson [1937], Wald and Wolfowitz [1944].

To learn more about Monte Carlo simulations and methods of computation in general, see Appendix 1.

## 3.10   Exercises

1. How was the analysis of the cell culture experiment described in Section 2.1 affected by the loss of two of the cultures due to contamination? Suppose these cultures had escaped contamination and given rise to the observations 90 and 95; what would be the results of a permutation analysis applied to the new, enlarged data set consisting of the following cell counts
   
   | Treated   | 121 | 118 | 110 | 90    |
   |-----------|-----|-----|-----|-------|
   | Untreated | 95  | 34  | 22  | 12 ?  |

2. In the preceding example, what would the result have been if you had used as your test statistic the difference between the sums of the first and second samples? the difference between their means? the sum of the squares of the observations in the first sample? the sum of their ranks?

3. Is my Acura getting less fuel efficient with each passing year? Use the data in Exercise 1.11 to test the null hypothesis against this alternative.

4. Cognitive dissonance theory suggests that when people experience conflicting motives or thoughts, they will try to reduce the dissonance by altering their perceptions. For example, college students were asked to perform a series of repetitive, boring tasks. They were then paid either $1 or $20 to tell the next student that the tasks were really interesting and fun. In private, they were asked to rate their own feelings about the tasks on a scale from 0 (dumb, dull, and duller) to 10 (exciting, better than sex).

   Those who received $20 for lying, assigned ratings of 3, 1, 2, 4, 0, 5, 4, 5, 1, 3. Those who received only $1, appeared to rate the tasks more favorably, 4, 8, 6, 9, 3, 6, 7, 10, 4, 8. Is the difference statistically significant? If you aren't a psychologist and would like to know what all this proves, see Festinger and Carlsmith [1959].

**5.** Babies can seem awfully dull for the first few weeks after birth. To us it appears that all they do is nurse and wet, nurse and wet. Yet in actuality, their brains are incredibly active. (Even as your brain, dulled by the need to wake up every few hours during the night to feed the kid, has gone temporarily on hold.) Your child is learning to see, to relate nervous impulses received from two disparate retina into concrete visual images. Consider the results of the following deceptively simple experiment in which an elaborate apparatus permits the experimenter to determine exactly where and for how long an infant spends in examining a triangle. (See Salaptek and Kessen, 1966, for more details and an interpretation.)

| Subject | Corners | Sides |
|---|---|---|
| Tim A. | 10 | 7 |
| Bill B. | 16 | 10 |
| Ken C. | 23 | 24 |
| Kath D. | 23 | 18 |
| Misha E. | 19 | 15 |
| Carl F. | 16 | 18 |
| Hillary G. | 12 | 11 |
| Holly H. | 18 | 14 |

**a.** Is there a statistically significant difference in viewing times between the sides and corners of the triangle?

**b.** Did you (and should you) use the same statistical test as you used to analyze the cognitive dissonance data?

**6.** Use Pitman correlation to test whether DDT residues have a deleterious effect on the thickness of a cormorant's egg shell.

| DDT residue in yolk (ppm) | 65 | 98 | 117 | 122 | 393 |
|---|---|---|---|---|---|
| Thickness of shell (mm) | 0.52 | 0.53 | 0.49 | 0.49 | 0.37 |

**7.** A group of mice were inoculated with Herpes virus type II. After 144 hours, the following virus titers were observed in their vaginas (see Good, 1979 for complete details):

| Saline controls | 10000, | 3000, | 2600, | 2400, | 1500. |
|---|---|---|---|---|---|
| Treated with antibiotic | 9000, | 1700, | 1100, | 360, | 1. |

**a.** Is this a one-sample, two-sample, $k$-sample, or matched pairs study?

**b.** Does treatment have an effect?

**c.** Most authorities would suggest using a logarithmic transformation before analyzing this data because of the exponential nature of viral growth (see Appendix 1). Repeat your analysis after taking the logarithm of each observation. Is there any difference? Compare your results and interpretations with those of Good [1979].

**8.** So you think you know baseball. Do home run hitters have the highest batting averages? Think about this hypothesis, then analyze the following experience based on a half season with the Braves.

| Batting average | .252 | .305 | .299 | .303 | .285 | .191 | .283 | .272 | .310 |
|---|---|---|---|---|---|---|---|---|---|
| Home runs | 12 | 6 | 4 | 15 | 2 | 2 | 16 | 6 | 8 |

| .266 | .215 | .211 | .244 | .320 |
|---|---|---|---|---|
| 10 | 0 | 3 | 6 | 7 |

If you don't believe this result, then check it out for your own favorite team.

9. In Exercise 12 of Chapter 2, you developed a theory based on the opinion data recorded in Exercise 3 of Chapter 1. Test that theory.

10. Construct a scatter plot for the law school data in Exercise 1.14, if you haven't already done so. What do you think is the nature of the association between undergraduate GPA and LSAT score? Is the correlation statistically significant?

11. To compare teaching methods, 20 school children were randomly assigned to one of two groups. The following are the test results:

| conventional | 65 | 79 | 90 | 75 | 61 | 85 | 98 | 80 | 97 | 75 |
|---|---|---|---|---|---|---|---|---|---|---|
| new | | 90 | 98 | 73 | 79 | 84 | 81 | 98 | 90 | 83 | 88 |

Are the two teaching methods equivalent in result?

12. To compare teaching methods, 10 school children were first taught by conventional methods, tested, and then taught by an entirely new approach. The following are the test results:

| conventional | 65 | 79 | 90 | 75 | 61 | 85 | 98 | 80 | 97 | 75 |
|---|---|---|---|---|---|---|---|---|---|---|
| new | | 90 | 98 | 73 | 79 | 84 | 81 | 98 | 90 | 83 | 88 |

Are the two teaching methods equivalent in result?
How does this experiment differ from that described in question 11?

13. Sketch a diagram to show if $G[x + 1] = F[x]$, where $F$, $G$ are two distribution functions, all the following are true: i) $G[x] < F[x]$ for all $x$, ii) $G$ is shifted to the right of $F$, iii) the median of $G$ is larger by 1 than the median of $F$.

14. You can't always test a hypothesis. A marketing manager would like to show that an intense media campaign just before Christmas resulted in increased sales. Should he compare this year's sales with last year's? What would the null hypothesis be? And what would be some of the alternatives?

15. How would you go about testing the hypothesis that the fuel additive described in Section 3.3.1 increases mileage by 10%?

16. Show the permutation distributions of the Pitman statistics $S_1$ with $g[i] = ai + b$ and $S_2$ with $g[i] = i$ are equivalent in the sense that if $S_1^\circ$ and $S_2^\circ$ are

the original values of these test statistics for a given set of observations, then $\Pr\{S_1 > S_1^\circ\} = \Pr\{S_2 > S_2^\circ\}$.

# CHAPTER 4

# When the Distribution Is Known

One of the strengths of the hypothesis-testing procedures described in the preceding chapter is that you need to know very little about the underlying population(s) to apply them. But suppose you have full knowledge of the processes that led to the observations in your sample(s), should you still use the same statistical tests? The answer is no, not always, particularly with very small or very large amounts of data. In this chapter, we consider several *parameter*-based distributions including the binomial (which you were introduced to in Chapter 2), the Poisson, and the normal or Gaussian, along with several other parametric distributions derived from them that are of value in testing location and dispersion.

## 4.1 Properties of Independent Observations

The *expected value* of an observation is the mean of all the values we would expect to observe were we to make a very large number of independent observations. In a population consisting of $N$ individuals, the expected value of an individual's height $h$ is the *population mean* $\frac{1}{N}\Sigma_{k=1}^{N} h_k$. It's easy to see that as a sample from this population grows larger, the sample mean $\frac{1}{n}\Sigma_{k=1}^{n} h_k$ must get closer to the population mean. If $X$ takes integer values $\{-N, \ldots, -2, -1, 0, +1, +2, \ldots, +M\}$, the expected value of $X$ is $\mu = \Sigma_{n=-N}^{M} n \Pr\{X = n\}$.

*The expected value* of the sum of two independent observations $X$ and $Y$ is equal to the expected value of $X$ plus the expected value of $Y$.

If $\mu$ is the expected value of $X$, the expected value of $(X - \mu)$ is 0 (see Exercise 9(a)).

If $\mu$ is the expected value of $X$, the *variance* of $X$ is defined as the expected value of $(X - \mu)^2$. In a population consisting of $N$ individuals, the variance of an individual's height is the *population variance* $\frac{1}{N}\Sigma_{k=1}^{N}(h_k - \mu)^2$. As a sample grows larger, the sample variance $s^2 = \Sigma_{i=1}^{n}(x_i - \overline{x})^2/(n-1)$ gets closer to the population variance.

If $X$ and $Y$ are the values taken by independent observations, then

1. expected value of $XY$ is expected value of $X$ times expected value of $Y$,

2. variance of $X + Y$ is the sum of the variance of $X$ and the variance of $Y$,

3. variance of $X - Y$ also is the sum of the variances of $X$ and $Y$.

The variance of the sum of $n$ independent observations is the sum of their variances. If the observations are identically distributed, each with variance $\sigma^2$, the sum of their variances is $n\sigma^2$. The variance of their mean $\bar{x}$ is $\sigma^2/n$ since

$$(\bar{x} - \mu)^2 = \left(\frac{1}{n}\sum_{k=1}^{n} x_k - \mu\right)^2 = \left(\frac{1}{n}\sum_{k=1}^{n}(x_k - \mu)\right)^2 = \frac{1}{n^2}\left(\sum_{k=1}^{n}(x_k - \mu)\right)^2.$$

Thus, the mean of 100 observations has 1/100th the variance of a single observation; the standard deviation of the mean of 100 observations, termed the *standard error* would have $1/\sqrt{100} = 1/10$th the standard deviation of a single observation. We'll use this latter property of the mean in Chapter 6 when we try to determine how large a sample should be.

## 4.2  Binomial Distribution

Recall that a binomial frequency distribution, written $B(n, p)$ results from a series of $n$ independent trials each with the same probability $p$ of success, and the same probability $1 - p$ of failure. The *mathematical expectation* or *expected value* of the number of successes in a single trial is $p$. This is the proportion of successes we would expect to observe in a very large number of repetitions of a single trial. If we perform a very large number $N$ of replications of $n$ binomial trials each with probability $p$ of success, we would expect to observe $np$ successes.

To determine the variance of the expected number of successes, note that we have two possible values for the squared deviations from the mean, a proportion $pn$ corresponding to successes that take the value $(1 - p)^2$, and a proportion $(1 - p)n$ corresponding to failures that take the value $(0 - p)^2$. A little algebra (Exercise 9) shows the variance is $np(1 - p)$.

### 4.2.1  Testing Hypotheses About a Binomial Distribution

Suppose we've flipped a coin in the air seven times, and six times it's come down heads. Do we have reason to suspect the coin is not fair? that $p$, the probability of throwing a head is greater than $1/2$?

To answer this question, we need to look at the frequency distribution of the binomial observation $X$ with $n$ independent trials and probability $p$ of success for each trial.

$$\Pr\{X = j\} = \binom{n}{j}p^j(1 - p)^{n-j} \quad \text{for } j = 0, 1, 2, \ldots, n.$$

If $n = 7$ and $j = 6$, this probability is $7p^6(1-p)$. For $p = 1/2$, this probability is $7/128 = 0.0547$ just slightly more than 5%. If six heads out of seven tries seems extreme to us, seven heads out of seven would have seemed even more extreme. Adding the probability of this more extreme event to what we have already, we see the probability of throwing six or *more* heads in seven tries is $8/128 = 0.0625$. While not significant at the 5% level, six or more heads out of seven tries does seem suspicious. If you were a mad scientist and observed that six times out of seven your assistant Igor began to bay like a wolf when there was a full moon, wouldn't you get suspicious?

## 4.3   Poisson: Events Rare in Time and Space

The decay of a radioactive element, an appointment to the United States Supreme Court, and a cavalry officer trampled by his horse have in common that they are relatively rare but inevitable events. They are inevitable, that is, if there are enough atoms, enough seconds or years in the observation period, and enough horses and momentarily careless men. Their frequency of occurrence has a Poisson distribution.

The number of events in a given interval has the Poisson distribution if it is the cumulative result of a large number of opportunities each of which has only a small chance of occurring. The interval can be in space as well as time. For example, if we seed a small number of cells into a Petri dish that is divided into a large number of squares, the distribution of cells per square follows the Poisson.[1]

If an observation $X$ has a Poisson distribution such that we may expect an average of $\lambda$ events per interval,

$$\Pr\{X = k\} = \frac{\lambda^j e^{-\lambda}}{j!} \quad \text{for } j = 0, 1, 2, \cdots.$$

An interesting and useful property of such observations is that the sum of a Poisson with expected value $\lambda_1$ and a second independent Poisson with expected value $\lambda_2$ is a Poisson with expected value $\lambda_1 + \lambda_2$.

### 4.3.1   Applying the Poisson

John Ross of the Wistar Institute held there were two approaches to biology: the analog and the digital. The analog was served by the scintillation counter: one ground up millions of cells then measured whatever was left behind in the stew; the digital was to be found in cloning experiments where any necessary measurements would be done on a cell-by-cell basis.

---

[1]The stars in the sky do not have a Poisson distribution, although it certainly looks this way. Stars occur in clusters, and these clusters, in turn, are distributed in superclusters. The centers of these superclusters do follow a Poisson distribution in space, and the stars in a cluster follow a Poisson distribution around the center almost as if someone had seeded the sky with stars and then watched the stars seed themselves in turn. See Neyman and Scott [1965].

John was a cloner and, later, as his student, so was I. We'd start out with 10 million or more cells in a 10 milliliter flask and try to dilute them down to one cell per milliliter. We were usually successful in cutting down the numbers to 10 thousand or so. Then came the hard part. We'd dilute the cells down a second time by a factor of 1:100 and hope we'd end up with 100 cells in the flask. Sometimes we did. Ninety percent of the time, we'd end up with between 90 and 110 cells, just as the binomial distribution predicted. But just because you cut a mixture in half (or a dozen, or a 100 parts) doesn't mean you're going to get equal numbers in each part. It means the probability of getting a particular cell is the same for all the parts. With large numbers of cells, things seem to even out. With small numbers, chance seems to predominate.

Things got worse when I went to seed the cells into culture dishes. These dishes, made of plastic, had a rectangular grid cut into their bottoms, so they were divided into approximately 100 equal size squares. Dropping 100 cells into the dish, meant an average of 1 cell per square. Unfortunately for cloning purposes, this average didn't mean much. Sometimes, 40% or more of the squares would contain two or more cells. It didn't take long to figure out why. Planted at random, the cells obey the Poisson distribution in space. An average of one cell per square means

$$\Pr\{\text{No cells per square}\} = 1 * e^{-1}/1 = 0.32$$

$$\Pr\{\text{One cell per square}\} = 1 * e^{-1}/1 = 0.32$$

$$\Pr\{\text{Two or more cells per square}\} = 1 - 0.32 - 0.32 = 0.36.$$

Two cells was one too many. A clone must begin with a single cell. I had to dilute the mixture a third time to ensure the percentage of squares that included two or more cells was vanishingly small. Alas, the vast majority of squares were now empty; I was forced to spend hundreds of additional hours peering through the microscope looking for the few squares that did include a clone.

## 4.3.2   Comparing Two Poissons

Suppose in designing a new nuclear submarine (or that unbuilt wonder, a nuclear spacecraft) you become concerned about the amount of radioactive exposure that will be received by the crew. You conduct a test of two possible shielding materials. During 10 minutes of exposure to a power plant using each material in turn as a shield, you record 14 counts with material A, and only four with experimental material B. Can you conclude that B is safer than A?

The answer lies not with the Poisson but the binomial. If the materials are equal in their shielding capabilities, then each of the 18 recorded counts is as likely to be obtained through the first material as through the second. In other words, under the null hypothesis you would be observing a binomial distribution with 18 trials each with probability $\frac{1}{2}$ of success or $B(18, \frac{1}{2})$. The numeric answer is left as an exercise (see Exercise 6).

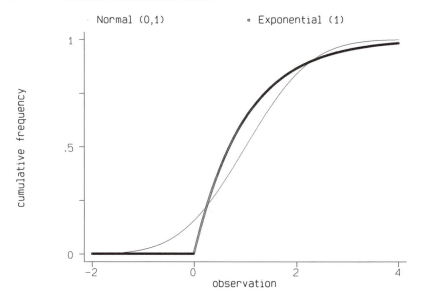

**Figure 4.1.** Cumulative distributions of normal and exponential observations with the same mean. The thin s-shaped curve is that of the normal or Gaussian; the thick curve is the exponential.

### 4.3.3   *Exponential Distribution*

If the number of events in a given interval has the Poisson distribution with $\lambda$ the expected number of events, then the time $t$ between events has the *exponential* distribution $F[t|\lambda] = 1 - e^{-\lambda t}$ for $t > 0$ that is depicted in Figure 4.1.

Now, imagine a system, one on a spacecraft, for example, where various critical components have been duplicated, so that $k$ consecutive failures are necessary before the system as a whole fails. If each component has a lifetime that follows the exponential distribution $F[t|\lambda]$, then the lifetime of the system as a whole obeys the *chi-square* distribution with $k$ degrees of freedom.[2]

## 4.4   Normal Distribution

Many times in our work we encounter observations $Y$, which may be explained by the following model:

$$Y = C + f[X_1, X_2, \ldots] + Z$$

where $C$ is a constant, $f[X_1, X_2, \ldots]$ is a function of certain explanatory variables $X_1$, $X_2$, ..., and $Z$ is a seemingly random amount, the sum of the effects of a large number of hard-to-pinpoint factors each of which makes only a small

---

[2]We'll encounter the chi-square distribution again in Sections 4.4.2 and 7.2.1.

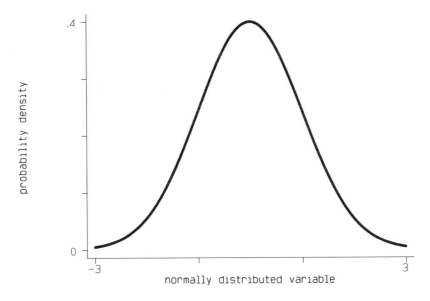

**Figure 4.2.** Probability distribution of a normally distributed variable.

contribution to the total. For example, the price you are willing to pay for an automobile depends upon the sticker price, how much you paid for your last car, and the prices of comparable cars, but it also depends to a varying degree on the salesperson's personality, your past experience with this brand and model of car, ads you've seen on TV, and how badly you need a new car.

In many, though not all cases, the frequency distribution of $Z$ takes the form depicted in Figure 4.2 where the unlabelled ticks on the $X$-axis represent one standard deviation below and one standard deviation above the population mode.

a. it is symmetrical about the mode;
b. its mean, median, and mode are the same;
c. most of its values, approximately 68%, lie within one standard deviation of the median;
d. about 95% lie within two standard deviations, yet arbitrarily large values are possible, even though with vanishingly small probability.

If we make repeated independent measurements (a classmate's height, the length of a flower petal), we will seldom get the same value twice (if we measure to sufficient decimal places) but rather a series of measurements that are normally distributed about a single central value.

An interesting and useful property of independent normally distributed observations is that their sum also has a normal distribution. In particular, if $\epsilon_1$ is normal with mean $\mu_1$ and variance $\sigma_1^2$, $N(\mu_1, \sigma_1^2)$, and $\epsilon_2$ is independent of $\epsilon_1$ and normal with mean $\mu_2$ and variance $\sigma_2^2$, $N(\mu_2, \sigma_2^2)$, then $\epsilon_1 + \epsilon_2$ is normal with mean $\mu_1 + \mu_2$ and variance $\sigma_1^2 + \sigma_2^2$.

## 4.4.1  Tests for Location

A parametric test for location based on tables of the normal distribution is seldom used in practice, since the normal $N(\mu, \sigma^2)$ involves not one but two parameters, and the population variance $\sigma^2$ is almost never known with certainty.

To circumvent this difficulty, W.E. Gosset, a statistician with Guinness Brewery, proposed the following statistics:

The test for $\mu = \mu_0$ for unknown variance is based on the distribution of

$$t = \frac{(\bar{x} - \mu_0)\sqrt{n}}{s} = \frac{(\bar{x} - \mu_0)\sqrt{n}}{\sqrt{\sum_{i=1}^{n}(x_i - \bar{x})^2/(n-1)}}$$

The test for $\mu_x = \mu_y$ for unknown but equal variances is based on the distribution of

$$t = \frac{(\bar{x} - \bar{y})}{\sqrt{\frac{(n_x-1)s_x^2+(n_y-1)s_y^2}{n_x+n_y-2}(1/n_x - 1/n_y)}}.$$

The first of these statistics for the one-sample case, has Student's $t$ distribution with $n - 1$ degrees of freedom; the second, used for the two-sample comparison, has Student's $t$ distribution with $n_x - n_y - 2$ degrees of freedom.[3] Today, of course, we don't need to know these formulae or to learn how to look up the $p$-values in tables; a computer will do the work for us with nearly all commercially available statistics packages. For samples of eight or more observations, the $t$ test generally yields equivalent results to the permutation tests introduced in Chapter 3. For smaller samples, the permutation test is recommended unless you can be absolutely sure the data are drawn from a normal distribution (see Section 6.4).

Sometimes, the changes observed in the data you collect are best expressed as percentages rather than absolute values. The best way to analyze such data is to first take the logarithms of such observations and then to apply either a $t$ test or the two-sample permutation comparison.

## 4.4.2  Tests for Scale

If an observation $x$ comes from a normal distribution with mean 0 and variance $\sigma^2$, then $x^2/\sigma^2$ has a *chi-square* distribution with one degree of freedom. Suppose we wish to test whether the actual variance is greater than or equal to $\sigma_0^2$ against the alternative it is less. We compare the observed value of $x^2/\sigma_0^2$ against a table of the chi-square distribution; if a value this large or larger would occur less than $\alpha$ of the time by chance, we reject the hypothesis; otherwise we accept it.

Most of the time, when we don't know what the variance is, we won't know what the mean is either. We can circumvent this problem by taking a sample of

---

[3]Why Student? Why not Gossett's test? Guinness did not want other breweries to guess it was using Gossett's statistical methods to improve beer quality, so Gossett wrote under a pseudonym as Student.

$n$ observations from the distribution and letting

$$S^2 = \frac{s^2}{\sigma_0^2} = \frac{\sum_{i=1}^{n}(x_i - \bar{x})^2/(n-1)}{\sigma_0^2}.$$

If the observations are from a normal distribution, our test statistic, which as you can see is based on the sample variance, will have the chi-square distribution with $n-1$ degrees of freedom. We can check its observed value against tables of this distribution to see if it is an extreme one. A major caveat is that the observations must be normal and, apart from observational errors, most real-world data are not (see, for example, Micceri, 1989).

Even small deviations from normality will render this approximation invalid My advice is to use one of the approaches described in Sections 3.3 and 3.6.1, whenever you need to test for scale. Use the chi-square distribution only when you have samples of 30 or more.

For comparing the variances of two normal populations, the test statistic of choice is

$$F = \frac{\sum_{i=1}^{n_x}(x_i - \bar{x})^2/(n_x - 1)}{\sum_{i=1}^{n_y}(y_i - \bar{x})^2/(n_y - 1)}.$$

A similar caveat applies. If you can't be sure the underlying distributions are normal, then don't use this test statistic, but adopt one of the procedures described in Chapter 8. Again, the exception is with large samples each of 30 or more in number; even then use the parametric approach only if you can be sure you are not working with mixtures of distributions.

## 4.5   Distributions

An observation $X$ is said to have the distribution $F$, if for all similar observations $F[x]$ is the probability that the observation takes on values less than or equal to $x$, that is $F[x] = \Pr\{X \le x\}$. $F$ is monotone nondecreasing in $x$, that is, it never decreases when $x$ increases, and $0 \le F[x] \le 1$.

The distribution function of the binomial $B(p, n)$ discussed in Section 4.2.1 is

$$F[k] = \sum_{j=0}^{k} \binom{n}{j} p^j (1-p)^{n-j} \quad \text{for } k = 0, 1, 2, \cdots, n.$$

The distribution function of the Poisson $P(\lambda)$ is

$$F[k] = \sum_{j=0}^{k} \lambda^j e^{-\lambda}/j! \quad \text{for } k = 0, 1, 2, \cdots.$$

Almost all random fluctuations have distributions that are made up of combinations or transformations of one or more of the normal, the binomial, and the

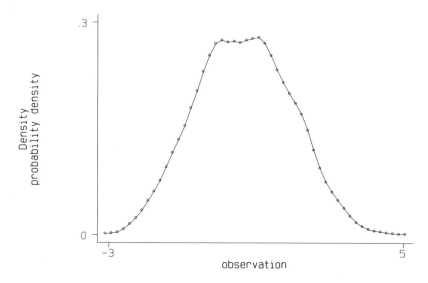

**Figure 4.3.** A sample from a mixture of two normal distributions.

Poisson. In most cases, it is difficult to characterize such mixed distributions by means of a formula. A simple example would be when a population is made up of two subpopulations, each of which has a normal distribution. Such a population would have a distribution that looks a great deal like Figure 4.3.

---

**Which Distribution?**

Counting the number of successes in $N$ independent trials? Binomial.
Counting the number of rare events that occur in a given time interval or a given region? Poisson.
Recording the length of the interval that elapses between rare events? Exponential.
What you observe is the sum of a large number of factors each of which makes only a small contribution to the total? Normal.

---

## 4.6   Applying What You've Learned

In this section, we attempt to apply some of the concepts we've learned in this and previous chapters. We begin by asking if the data come from one of the distributions we've just been introduced to.

Suppose you've got this strange notion that your college's hockey team is better than mine. We compare win/lost records for last season and see that while McGill won 11 of its 15 games, your team only won 8 of 14. But is this difference statistically significant?

With the outcome of each game being success or failure, and successive games being independent of one another, it looks at first glance as if we have two series of binomial trials (as we'll see in a moment, this is highly questionable). Using the methods described in Section 5.2, we could derive confidence intervals for each of the two binomial parameters. If these intervals do not overlap, then the difference in win/loss records is statistically significant. Fisher's Exact Test, described in Section 7.1, is an even better way to compare two binomials

"Unfair," you cry, "we haven't got to Chapters 5 and 7 yet." Okay; let's make the comparison another way by comparing total goals. McGill scored a total of 28 goals last season and your team 32. Using the approach described in Section 4.4.2, we could look at this set of observations as a binomial with $28 + 32 = 60$ trials, and test the hypothesis that $p \leq 1/2$ (that is, McGill is no more likely to have scored the goal than your team) against the alternative that $p > 1/2$.

This latter approach has several problems. For one, your team played fewer games than McGill. But more telling, and the principal objection to all the methods we've discussed so far, the schedules of our two teams may not be comparable.

With binomial trials, the probability of success must be the same for each trial. Clearly, this is not the case here. We need to correct for the differences among opponents. After much discussion—what else is the off-season for?—you and I decide to award points for each game using the formula $S = O + GF - GA$, where GF stands for goals for, GA for goals against, and O is the value awarded for playing a specific opponent. In coming up with this formula and with the various values for O, we relied not on our knowledge of statistics but on our hockey expertise. This reliance on domain expertise is typical of most real-world applications of statistics.

The point totals we came up with read like this

| McGill | $4, -2, 1, 3, 5, 5, 0, -1, 6, 2, 2, 3, -2, -1, 4$ |
| Your School | $3, 4, 4, -3, 3, 2, 2, 2, 4, 5, 1, -2, 2, 1$ |

Curiously, your school's first four point totals, all involving games against teams from other leagues, were actually losses, their high point value being the result of the high-caliber of the opponents. I'll give you guys credit for trying.

We can analyze these observations in one of two ways: i) using the two-sample comparison described in Chapter 2, or ii) using the $t$ test described in Section 4.3. Arguing against the $t$ test is that the observations come from a mixture of populations (league games and games against non-league opponents). The permutation test is recommended and the comparison is left to you as an exercise.

## 4.7    Summary and Further Readings

In this chapter, you had a brief introduction to parametric distributions including the binomial, Poisson, Gaussian, and exponential, and to several distributions derived from them including $\chi^2$, Students $t$, and $F$. Relationships among these and other common univariate distributions are described in Lemmis [1986]. Mixtures of these distributions are discussed in McLachlan and Peel [2000]. You learned here how to obtain tests of hypothesis based on these distributions. For a further discussion of the underlying theory, see Lehmann [1986]. *The CRC Handbook of Tables for Probability and Statistics* can help you in deriving the appropriate cutoff values.

Those with an interest in ranking sports teams will enjoy the article by John O'Keefe in *Sports Illustrated*, May 7, 2001.

## 4.8   Exercises

1. Seventy percent of the registered voters in Orange County are Republican. What is the probability the next two people in Orange County you ask about their political preferences will say they are Republican? The next two out of three people? What is the probability that at most one of the next three people will admit to being a Democrat?

2. Two thousand bottles of aspirin have just rolled off the assembly line, 15 have ill-fitting caps, 50 have holes in their protective covers, and five are defective for both reasons. What is the probability that a bottle selected at random will be free from defects? What is the probability that a group of 10 bottles selected at random will all be free of defects?

3. For each of the following indicate whether the observation is binomial, Poisson, exponential, normal or almost normal, comes from some other distribution, or is predetermined (that is, not random at all).

   a. Number of books in your local public library.

   b. Guesses as to the number of books in your local public library.

   c. Heights of Swedish adults.

   d. Weights of Norwegian women.

   e. Alligators in an acre of Florida swampland.

   f. Vodka drinkers in a sample of 800 Russians.

   g. Messages on your answer machine.

   h. You shoot at a target: the distance from the point of impact to the target's center.

4. How many students in your class have the same birthday as the instructor? In what proportion of classes, all of size 20, would you expect to find a student who has the same birthday as the instructor? two students who have the same birthday?

5. What is the largest number of cells you can drop into a Petri dish divided into 100 squares and be sure the probability a square contains two or more cells is less than 1%? What percentage of the squares would you expect would contain exactly one cell?

6. Given a binomial distribution of 18 trials each with probability 1/2 of success. What is the probability of observing 14 or more successes?

7. Bus or subway? The subway is faster, but there are more buses on the route. One day, you and a companion count the number of arrivals at the various terminals. Fourteen buses arrive in the same period as only four subway trains pull in. As in Section 4.3.2, use the binomial distribution to determine whether the bus is a significantly better choice than the subway.

8. Reanalyze the data of Exercises 11 and 12 assuming the data in these exercises came from a normal distribution.

9. $X$, $Y$ are independent observations from the same distribution taking integer values $\{\ldots, -2, -1, 0, +1, +2, \ldots\}$. If $\mu$ is the expected value of $X$, that is, $\mu = \Sigma_{n=-\infty}^{\infty} n \, Pr\{X = n\}$, we write $E(X) = \mu$. Show that
   a. the expected value of $(X - \mu)$ is 0.
   b. $E(XY) = E(X)E(Y)$.
   c. the variance of $X - Y$ is the sum of the variances of $X$ and $Y$.

10. Show that the variance of a binomial with $n$ trials and probability $p$ of success of each trial is $np(1 - p)$.

11. Suppose $X$ comes from a distribution for which $Pr\{x = 0\} = 1/5$, $Pr\{x = 1\} = 1/5$, $Pr\{x = 2\} = 1/5$, $Pr\{x = 3\} = 1/5$, $Pr\{x = 4\} = 1/5$ and $Y$ comes from a distribution for which $Pr\{y = 0\} = 1/10$, $Pr\{y = 1\} = 1/5$, $Pr\{y = 2\} = 1/5$, $Pr\{y = 3\} = 1/5$, $Pr\{y = 4\} = 3/10$. Show that the expected value of $Y$ is greater than the expected value of $X$.

12. A poll taken in January 1998 of 1000 people in the United States revealed that 64% felt President Clinton should remain in office despite certain indiscretions. What would you estimate the variance of this estimate to be? Suppose 51% is the true value of the proportion supporting Clinton, what would the true value of the variance of your original estimate be?

# CHAPTER 5

---

# Estimation

In this chapter, you expand on the knowledge of estimation gained in Chapter 1 to make point and interval estimates of the effects you detected in Chapter 3.

## 5.1  Point Estimation

"Just give me the bottom line," demands your supervisor, "one number I can show my boss." The recommended point or single-value estimate is the *plug-in* estimate in most cases: For the population mean, use the sample mean; for the population median, use the sample median; and to estimate a proportion in the population, use the proportion in the sample. These plug-in estimates have the virtue that are *consistent*, that is, as the sample grows larger and larger, the plug-in estimate grows closer and closer to the true value. In many instances, particularly for measures of location, plug-in estimates are *unbiased*, that is, they are closer on the average to the true value than to any other value.

The exception that proves the rule is the population variance that you consider in Exercise 6. The recommended estimate is $\frac{1}{n-1}\sum_{i=1}^{n}(x_i - \bar{x})^2$ while the plug-in estimate is $\frac{1}{n}\sum_{i=1}^{n}(x_i - \bar{x})^2$.

The recommended estimate is recommended because if we were to take a large number of samples, each of size $n$, the mean or mathematical expectation of the estimate would be the population variance. While both estimates are consistent, only the recommended estimate is unbiased.

It is customary, though not always advisable, when reporting results to provide both the estimate and an estimate of its standard error. The bootstrap will help us here. We pretend the sample is the population and take a series of bootstrap samples from it. We make an estimate for each one of these samples; label this estimate $\hat{\theta}_b$. After we have taken $B$ bootstrap samples, we compute the standard error of our sample of $B$ estimates

$$\sqrt{\frac{1}{B-1}\Sigma_{b=1}^{B}(\hat{\theta}_b - \bar{\hat{\theta}})^2},$$

where $\bar{\bar{\theta}}$ is the mean of the bootstrap estimates.

Suppose we want to make an estimate of the median height of Southern California sixth-graders based on the heights of my sixth-grade class. Our estimate is simply the median of the class, 153.5 cm. To estimate the standard error or precision of this estimate, we take a series of bootstrap samples from the heights recorded in Table 1.1. After reordering each sample from smallest to largest, we have

Bootstrap Sample 1: 137.0 138.5 138.5 141.0 142.0 145.0 145.0 147.0 148.5 148.5 150.0 150.0 154.0 155.0 156.5 157.0 158.0 158.5 159.0 160.5 161.0 161.0 Median = 150.0

Bootstrap Sample 2: 138.5 140.0 140.0 141.0 142.0 143.5 145.0 148.5 150.0 150.0 153.0 154.0 156.5 157.0 158.0 158.5 159.0 160.5 161.0 162.0 162.0 167.5 Median = 153.5

Bootstrap Sample 3: 140.0 141.0 142.0 143.5 145.0 147.0 148.5 150.0 153.0 155.0 156.5 156.5 157.0 158.0 158.5 159.0 159.0 159.0 160.5 161.0 162.0 167.5 Median = 156.5

If we were to stop at this point (normally, we'd take at least 50 bootstrap samples), our estimate of the standard error of our estimate of the median would be

$$\sqrt{\frac{1}{(3-1)}\{(150-152.8)^2 + (151.5-152.8)^2 + (155.75-152.8)^2\}}.$$

## 5.2  Interval Estimation

The standard error is only an approximation to the actual dispersion of an estimate. If the observations are normally distributed (see Section 4.4) or if the sample is very large, then the interval from one standard error below the sample mean to one standard error above it will cover the true mean of the population about two-thirds of the time. But many estimates (for example, that of a variance or a correlation coefficient) do not have a symmetric distribution so that such an interpretation of the standard error would be misleading. In the balance of this section, you learn to derive a more accurate interval estimate, one, hopefully, more likely to cover the true value than a false one.

### 5.2.1  Detecting an Unfair Coin

We begin our investigation of improved interval estimates with the simplest possible example, a binomial variable.

My friend Tom settles all his arguments by flipping a coin. So far today, he's flipped five times, called heads five times, and got his own way five times. My

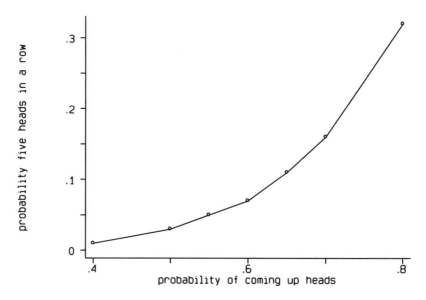

**Figure 5.1.**   Deriving lower confidence bounds for the binomial.

natural distrust of my fellow humans leads me to suspect Tom's not using a fair coin. Recall from Section 2.5 that if $p$ is the probability of throwing a head,

$$\Pr\{5 \text{ heads in a row} \mid \text{for probability of success } p\} = p^5(1-p)^0.$$

If $p = 0.5$ or 50% as it would for a fair coin, this probability is 0.032. Awfully small. If $p = 0.65$, this probability is 11%. Could be the coin is weighted so that $p = 0.65$. In fact, for any value of $p > 0.55$, the probability of tossing five heads in a row is at least 5%. But not for $p = 0.5$. Do you suppose Tom's coin is weighted so it comes up heads at least 55% of the time? No. He wouldn't do a thing like that. He's my friend.

In order to be able to find a *lower confidence bound* on $p$ for any desired confidence level, I built the chart shown in Figure 5.1. To determine a lower bound, first find the complement of the confidence level on the $y$-axis; then read across to the curve and down to the $x$-axis. For example, if the desired confidence level is 95% or 0.95, locate $1-0.95 = .05$ on the $y$-axis, then read across to the curve and down to the $x$-axis to find $p > 0.55$. If you are willing to be less confident, say you'd settle for being right 90% of the time, the corresponding lower confidence bound is about 0.64.

This chart is applicable not only to suspect coins but to any example involving variables that have exactly two possible outcomes.

## 5.2.2   Confidence Bounds for a Sample Median

In this next example, we've a set of observations $-1, 2, 3, 1.1, 5$ taken from a much larger population and want to find a *confidence interval* that will cover the true value of the unknown median $\theta$ in $1 - \alpha$ percent of similar instances in which we take samples and form estimates. As an alternative to the bootstrap described in Section 1.6.4, the method of randomization can help us find an interval estimate that has this property.

Recall from Section 3.6.2 that in the first step of our permutation test for the location parameter of a single sample, H: $\theta = \theta_o$ versus K: $\theta > \theta_o$, we subtract $\theta_o$ from each of the observations. We might test a whole series of hypotheses involving smaller and smaller values for $\theta_o$ until we find a $\theta_1$ such that as long as $\theta_o \geq \theta_1$, we accept the hypothesis at the $\alpha$ significance level, but if $\theta_o < \theta_1$ we reject it. Then a $1 - \alpha$ confidence interval for $\theta$ is given by the one-sided interval $\{\theta \geq \theta_1\}$.

Let's begin by testing the hypothesis $\theta = 0$. Our first step is to compute the sum of the negative deviations about 0, which is 1.[1] Under the null hypothesis, this distribution is symmetric about its mean of zero. Positive and negative values are equally likely. We would be just as likely to observe a "+1" as a "−1" or a "−2" as a "+2." Among the $2 \times 2 \times 2 \times 2 \times 2 = 2^5 = 32$ possible reassignments of plus and minus signs are

$$+1, -2, +3, +1.1, +5$$
$$+1, +2, +3, +1.1, +5$$

and

$$-1, -2, +3, +1.1, +5.$$

Our next step is to compute the sum of negative deviations for each rearrangement. For the three arrangements shown above, this sum would be 2, 0, and 3, respectively. Only two of the 32 possible rearrangements, or $1/16^{th}$, result in samples as extreme as the original observations. We accept the hypothesis that $\theta = 0$ at the 1/16th level and thus at any smaller $\alpha$-level including the $1/32^{nd}$. Similarly, we accept the hypothesis that $\theta = -0.5$ at the 1/32nd level, and even that $\theta = -1 + \varepsilon$, where $\varepsilon$ is an arbitrarily small but still positive number. But we would reject the hypothesis that $\theta = -1$ as after subtracting $-1$ the transformed observations 0, 2, 3, 1.1, 5 are all as large or larger than zero.

Our one-sided confidence interval is $\{\theta > -1\}$, and we have confidence that $31/32^{nd}$ of the time this method yields an interval that includes the true value of the location parameter. It is the method that inspires this degree of confidence: Other samples can and probably will yield different results.[2]

Our one-sided test of a hypothesis gives rise to a one-sided confidence interval. But knowing that $\theta$ is larger than $-1$ may not be enough. We may want to pin it down to a more precise two-sided interval, that $\theta$ lies between $-1$ and $1$, for instance.

---

[1] I chose negative deviations rather than positive as there are fewer of them and thus fewer calculations, at least initially.

[2] See Section 6.1.3 for a discussion of the relationships among sample size, interval length, and confidence.

To obtain a two-sided interval, we need to begin with a two-sided test. Our hypothesis for this test is that $\theta = \theta_o$ against the two-sided alternative that $\theta$ is either smaller or larger than $\theta = \theta_o$. We use the same test statistic, the sum of the negative observations, that we used in the previous one-sided test. Again, we look at the distribution of our test statistic over all possible assignments of the plus and minus signs to the observations. But this time we reject the hypothesis if the value of the statistic for the original observations is either one of the largest or one of the smallest of the possible values.

In our example, we don't have enough observations to find a two-sided confidence interval at the 31/32nd level, so we'll try to find one at the 15/16ths. The lower boundary of the new confidence interval is still $-1$. But what is the new upper boundary? If we subtract 5 from every observation, we would have the values $-6, -4, -2, -3.9, 0$; their sum is $-15.9$. Only the current assignment of signs to the transformed values, that is, only one out of the 32 possible assignments, yields this large a sum for the negative values. The symmetry of the two-sided permutation test requires that we set aside another 1/32nd of the arrangements at the high end. Thus we would reject the hypothesis that $\theta = 5$ at the $1/32 + 1/32$ or 1/16th level. Consequently, the chances are 15/16 that the interval $\{-1, 5\}$ covers the unknown parameter value.

---

**Lower Confidence Bound for a Median:**
**Permutation Approach**

Guess at a Lower Bound L.
Test the Hypothesis the Median is L.
    If you accept this hypothesis, decrease the bound and test again.
    If you reject this hypothesis, increase the bound and test again.

---

These results are readily extended to a confidence interval for the *vector* of parameters that underlies a one-sample, two-sample, or $k$-sample experimental design with single- or vector-valued variables. For example, we might want a simultaneous estimate of the medians of both arm span and height. The $1 - \alpha$ confidence interval consists of all values of the parameter vector for which we would accept a hypothesis concerning the two medians at the $\alpha$ significance level. Remember, one-sided tests produce one-sided intervals and two-sided tests produce two-sided confidence intervals.

## 5.2.3   Confidence Intervals and Rejection Regions

There is a direct connection between confidence intervals and the rejection regions of our tests. Suppose $A(\theta')$ is an $1 - \alpha$ level acceptance region for testing the hypothesis $\theta = \theta'$, that is, we accept the hypothesis if our test statistic $T$ belongs to the acceptance region $A(\theta')$ and reject it otherwise. Suppose $S(X)$ is an $1-\alpha$ level confidence interval for $\theta$ based on the set of observations $X = \{x_1, x_2, \ldots, x_n\}$.

Then, $S(X)$ consists of all the parameter values $\theta^*$ for which $T[X]$ belongs to the acceptance region $A(\theta^*)$.

$$\Pr\{S(X) \text{ includes } \theta_o \text{ when } \theta = \theta_o\} = \Pr\{T[X] \in A(\theta_o) \text{ when } \theta = \theta_o\} \geq 1 - \alpha.$$

Suppose our hypothesis is $\theta = 0$ and we observe $X$. Then we accept this null hypothesis if and only if our confidence interval $S(X)$ includes the value 0.

---

**Important Terms**

*Acceptance Region*, $A(\theta_o)$. Set of values of the statistic $T[X]$ for which we would accept the hypothesis H: $\theta = \theta_o$. Its complement is called the rejection region.

*Confidence Region*, $S(X)$. Also referred to as a confidence interval (for a single parameter) or a confidence ellipse (for multiple parameters). Set of values of the parameter $\theta$ for which given the set of observations $X = \{x_1, x_2, \ldots, x_n\}$ and the statistic $T[X]$ we would accept the corresponding hypothesis.

---

To illustrate the duality between confidence bounds and rejection regions, we've built the curve shown in Figure 5.2. (See Exercise 3 for the details of the construction.) Now, let's take a second look at the problem of the unfair coin that we considered at the beginning of this chapter. Only instead of tossing a coin, let's consider a survey in which we ask five individuals selected at random whether they plan to vote for the Green candidate or not. We use the number of positive responses in the five interviews to obtain a lower confidence bound for the probability $p$ that an individual selected from the population at large will vote for the Green candidate. Turning to Figure 5.2, and entering the chart from below along the horizontal axis, we see that if 2 of the 5 favor the Green, a 95% lower confidence bound for $p$ is 0.09. Alternatively, if we wish to test the hypothesis that $p = 0.20$, we enter the chart from the left side along the vertical axis to learn we should reject at the 5% level if three or more of the selected individuals favor Green.

## 5.2.4    One-Sample Test for the Variance

We can apply the results of 5.2.3 immediately to obtain a one-sample test for the variance $\sigma^2$ by using the bootstrap to obtain a confidence interval for the variance. If this confidence interval contains the hypothesized value $\sigma_0^2$, accept the hypothesis; reject it otherwise. To improve the accuracy of the confidence interval and increase the probability of rejecting the hypothesis if it is untrue, apply the methods of Section 5.4.

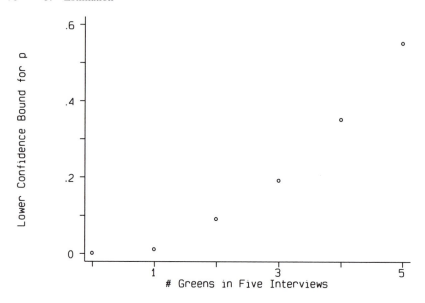

**Figure 5.2.** Lower confidence bound as a function of the number of positive responses in five binomial trials.

## 5.2.5   *Behrens-Fisher Problem*

A major distinction between the permutation test and the bootstrap is that the former is based on the equivalence of certain *distributions*, while the latter is based on interval estimates of *parameters*. Permutation methods cannot be used to test whether the means of two populations are the same *unless* the variances of the two populations are also the same. The bootstrap does not have this restriction. We can obtain a bootstrap test from a confidence interval based on the difference of the two sample means. A more accurate result—one more likely to include true values of the difference and exclude false ones as shown by Hall and Wilson [1991], can be obtained from the confidence interval for the ratio

$$\frac{(\bar{x} - \bar{y})}{\sqrt{\sum(x - \bar{x})^2 + \sum(y - \bar{y})^2}}.$$

To obtain a confidence interval for this ratio, we bootstrap separately from each of the samples each time.

## 5.3   Bivariate Correlation

Although we may perform an initial test as we did in Section 3.5 to determine whether two variables $X$ and $Y$ are correlated, our primary concern is with the proportion of the observation $Y$ that is explained by $X$. One measure of this

Table 5.1. Test Results from Mardia, Kent, and Bibby[1979]

| Algebra | 67 | 80 | 71 | 63 | 65 | 72 | 65 | 68 | 58 | 60 | 60 | 59 |
|---------|----|----|----|----|----|----|----|----|----|----|----|----|
| Statistics | 81 | 81 | 81 | 68 | 63 | 73 | 68 | 56 | 70 | 45 | 54 | 44 |

proportion and the degree of association between $X$ and $Y$ is the correlation $\rho$ defined as

$$\rho = \frac{Cov(XY)}{\sigma_X \sigma_Y}.$$

The covariance when two variables $X$ and $Y$ are observed simultaneously is

$$Cov(XY) = \Sigma_{i=1}^{n}(x - \bar{x})(y - \bar{y})/(n - 1).$$

If $X$ and $Y$ are independent and uncorrelated then $\rho = 0$. If $X$ and $Y$ are totally and positively dependent, for example, if $X = 2Y$, then $\rho = 1$. In most cases of dependence $0 < |\rho| < 1$.

If we knew the standard deviations of $X$ and $Y$, we could use them to help obtain a confidence interval for $\rho$ by the method of permutations. But we only know their estimates.

.1329749                                rho                                .9635068

**Figure 5.3.**    Results of 400 bootstrap simulations using Table 5.1.

Fortunately, we can use the bootstrap to obtain a confidence interval for $\rho$ in a single step, just as we did in Section 1.6.4. Does a knowledge of algebra ensure a better grade in statistics? Table 5.1 displays test score data from a study by Mardia, Kent, and Bibby [1979]. A box plot of the bootstrap results using a Stata routine (see Appendix 3) is depicted in Figure 5.3. The median correlation is 0.68, with a 90% confidence interval of {0.50, 0.87}. As this interval does not include 0, we reject the null hypothesis and conclude that a knowledge of algebra does indeed ensure a better grade in statistics (see Section 5.2.2). See, also, Young [1988].

---

**Confidence Interval for the Correlation Coefficient**

Derive the bootstrap distribution:
  Choose a bootstrap sample without replacement.
  Compute the correlation coefficient.
  Repeat several hundred times. Select a confidence interval from this distribution.

## 5.4    Improving the Bootstrap Estimates

The percentile bootstrap interval described in Sections 1.6 and depicted in Figure 5.3 is notoriously *biased* and inaccurate. A "95%" confidence interval may only cover the correct parameter value 90% of the time. Worse, it may cover an incorrect value of the parameter more often than it covers the correct one. In the following sections we describe three methods for increasing its accuracy.

### *5.4.1    Bias-Corrected Bootstrap*

The bias-corrected and accelerated $BC_a$ interval represents a substantial improvement for one-sided confidence intervals, though for samples under size 20, it is still suspect. The idea behind these intervals comes from the observation that percentile bootstrap intervals are most accurate when the estimate is symmetrically distributed about the true value of the parameter and the tails of the estimate's distribution drop off rapidly to zero. The normal distribution depicted in Figure 4.2 represents this ideal.

Suppose $\theta$ is the parameter we are trying to estimate, $\hat{\theta}$ is the estimate, and we are able to come up with a monotone increasing transformation $m$ such that $m(\theta)$ is normally distributed about $m(\hat{\theta})$. We could use this normal distribution to obtain an unbiased confidence interval, then apply a back-transformation to obtain an almost-unbiased confidence interval.

The method is not as complicated as it reads because we don't actually have to go through all these steps, merely agree that we could if we needed to. The resulting formula *is* complicated.[3] Fortunately, S code for obtaining $BC_a$ intervals may be obtained from the statistics archive of Carnegie-Mellon University by sending an email to statlib@lib.stat.cmu with the one-line message "send bootstrap.funs from S." A SAS macro is available at the following URL: www.asu.edu/it/fyi/research/helpdocs/statistics/SAS/tips/jackboot.html.

The latest version of Stata also provides for bias-corrected intervals via its .bstrap command. Applying this command to the height data from my classroom, we obtain the following results:

```
.using class dta
. bs ''summarize height,detail'' ''r(p50)'',reps(50) saving(bclass)

command: summarize height,detail
statistic: r(p50); the median, (obs=22)

Bootstrap statistics
Reps Observed Bias  Std. Err.    [95% Conf. Interval]
--------------------------------------------------------
50    153.5   -.41   3.1142      147. 157.5 (P)
                                 145. 157.5 (BC)
--------------------------------------------------------
              P = percentile, BC = bias-corrected
```

---

[3] See Chapter 14 of Efron and Tibshirani, 1993.

This is the exception that proves the rule, for in this example the bias-corrected interval (BC) is *wider* than the percentile interval (P), thus resulting in an increase in the Type II error. Stata's bstrap routine also provides an estimate of the possible bias resulting from the use of the sample mean in place of the population mean. This estimate should NOT be used to correct the sample mean. Instead, we may use the "optimism" described in Section 10.5.

Applying the bias correction to the process temperature data of Table 3.2, the resultant interval is both shorter than the percentile interval and shifted to the right.

```
. bs ''summarize var1,meanonly'' ''r(mean)'',reps(100)}

command: summarize var1,meanonly}
statistic: r(mean); (obs=20)}

Bootstrap statistics}

Reps Observed Bias Std. Err. [95% Conf. Interval]
-----------------------------------------------------
100 454.55 -.412 3.908 446.25 461.05 (P)}
447.15 461.35 (BC)}
-----------------------------------------------------
P = percentile, BC = bias-corrected
```

## 5.4.2  Smoothing the Bootstrap

If you haven't kept up your math, the balance of this section can be challenging. Two alternatives suggest themselves: 1) read through the next few sections quickly to get a feel for the type of problems that can be solved and come back to them if and when you need them; or 2) work through the algebra step by step substituting real data for symbols.

An inherent drawback of the bootstrap, particularly with small samples, lies in the discontinuous nature of the empirical distribution function. Presumably, our samples come from a continuous or near-continuous distribution. Figure 5.4 illustrates the distinction. The jagged curve is the empirical distribution function. The smooth curve that passes through it was obtained by replacing the original observations with their Studentized[4] equivalents, $z_i = (x_i - \bar{x})/s$, then plotting the distribution function of a normally distributed random variable $N(\bar{x}, s^2)$.

To obtain an improved bootstrap confidence interval for the median height of a typical sixth-grader based on the data in Table 1.1, we modify our bootstrap procedure as follows:

1. Sample $z_1, z_2, \ldots, z_n$ with replacement from the original observations $x_1$, $x_2, \ldots, x_n$.

---

[4]So-called because of its similarity in form to Student's $t$.

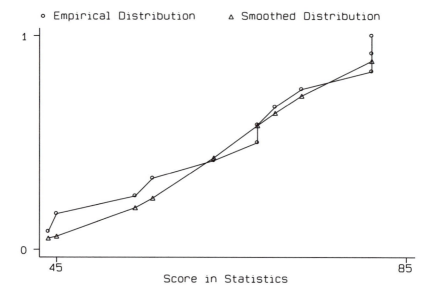

**Figure 5.4.** Cumulative distribution of statistics scores from Table 5.1. The jagged curve o–o–o is the empirical (observed) distribution function.

2. Compute their mean $\bar{z}$ and the plug-in estimate of the population variance

$$\hat{\sigma}^2 = \sum_{i=1}^{n} (z_i - \bar{z})^2 / n.$$

3. Set

$$x_i^* = \bar{z} + \frac{(z_i - \bar{z} + h\varepsilon_i)}{\sqrt{1 + h^2/\hat{\sigma}^2}},$$

where the $\varepsilon_i$ are drawn from an $N(0, 1)$ distribution, and $h$ is a smoothing constant, chosen by trial and error.

4. Use $x_1^*, x_2^*, \ldots, x_n^*$ to estimate the median.

To obtain a smoothed bootstrap estimate of the distribution of the correlation coefficient $\rho$, we use a similar procedure:

1. We take a bootstrap sample $(x, y)_1^*, (x, y)_2^*, \ldots, (x, y)_n^*$ with replacement from the original set of simultaneous observations on two *variables* $(x, y)_1$, $(x, y)_2, \ldots, (x, y)_n$.

2. Following Hall and Wilson [1991], we compute the mean $(\bar{x}^*, \bar{y}^*)$ of the bootstrap sample, and its covariance matrix $\Sigma^*$.

3. We draw $n$ pairs $(\varepsilon_i, \eta_i)$ from the bivariate normal distribution with mean $(\bar{x}^*, \bar{y}^*)$ and covariance matrix $\Sigma^*$, and use them to form

$$x_i^* = \bar{x}^* + \frac{(x_i - h\varepsilon_i)}{\sqrt{1 + h^2/\sigma_x^{2*}}}, \quad y_i^* = \bar{y}^* + \frac{(y_i - h\varepsilon_i)}{\sqrt{1 + h^2/\sigma_x^{2*}}}.$$

4. Finally, we use $(x_1^*, y_1^*), (x_2^*, y_2^*), \ldots, (x_n^*, y_n^*)$ to estimate the correlation coefficient.

---

**Computing a Smoothed Bootstrap Estimate of the Median Sample Calculation with Stata**

. bsample _N

. list height
143.5 160.5 155 137 156.5 158
161 148.5 153 141 145 140
147 154 138.5 159 155 158
141 141 145 138.5
$\bar{z} = 149$ and $\hat{\sigma}^2 = 64.18$
. invnorm (uniform)
If we take $h = 0.5$, then

$$x_1^* = 148.9 + \frac{143.5 - 148.9 + 0.5\varepsilon_1}{\sqrt{1 + 0.25/64.18}} = 143.54$$

---

## 5.4.3   Iterated Bootstrap

Bootstrap iteration provides another way to improve the accuracy of bootstrap confidence intervals. Let $I$ be a $1 - \alpha$ level confidence interval for $\theta$ whose actual coverage $\pi_\theta[\alpha]$, depends upon both the true value of $\theta$ and the hoped-for cover age $1 - \alpha$. In most cases $\pi_\theta[\alpha]$ will be larger than $1 - \alpha$. Let $\alpha'$ be the value of $\alpha$ for which the corresponding confidence interval $I'$ has probability $1 - \alpha$ of covering the true parameter. We can't solve for $\alpha'$ and $I'$ directly. But we can obtain a somewhat more accurate confidence interval $I*$ based on $\alpha*$ the estimate of $\alpha'$ obtained by replacing $\theta$ with its plug-in estimate and the original sample with a bootstrap sample. Details of the procedure are given in Martin [1990, pp. 113-114]. The iterated bootstrap, while straightforward, is computationally intensive. Suppose the original sample has 30 observations, and we take 300 bootstrap samples from each of 300 bootstrap samples. That's 2,700,000 values, plus an equal number of calculations![5]

## 5.5   Summary

In this chapter, you learned how to make point estimates and to estimate their precision using the bootstrap. You learned how to make interval estimates using

---

[5] A third iteration involving some 800 million numbers will further improve the interval's accuracy.

either the bootstrap or the permutation test. You studied the relationship between confidence intervals and acceptance regions and learned how to improve bootstrap estimates.

## 5.6   To Learn More

For further information on deriving confidence intervals using the randomization approach see Lehmann [1986, pp. 246–263], Gabriel and Hsu [1983], Garthwaite [1996], John and Robinson [1983], Maritz [1983, p. 7, p. 25], Martin-Lof [1974], Noether [1978], Tritchler [1984], Buonaccorsi [1987]. Solomon [1986] and Good [2001] consider the application of confidence intervals in legal settings. For a discussion of the strengths and weaknesses of pivotal quantities, see Berger and Wolpert [1984].

For examples of the wide applicability of the bootstrap method, see Chernick [1999] and Chapter 7 of Efron and Tibshirani [1993]. These latter authors comment on when and when not to use bias-corrected estimates. The early literature relied erroneously on estimates of the mean and covariance derived from the original rather than the bootstrap sample. The improved method used today is due to Hall and Wilson [1991]; see guidelines for bootstrap testing in Section 6.3. Other general texts and articles of interest include Efron and Tibshirani [1986, 1991], Mooney and Duval [1993], Stine [1990], and Young [1994]. DiCiccio and Romano [1988] and Efron and DiCiccio [1992] discuss bootstrap confidence intervals.

The smoothed bootstrap was introduced by Efron [1979] and Silverman [1981], and is considered at length in Silverman and Young [1987], Praska Rao [1983], DiCiccio, Hall, and Romano [1989], and Falk and Reiss [1989]. For discrete distributions, the approach of Lahiri [1993] is recommended. Hårdle [1991] provides S-Plus computer algorithms. The weighted bootstrap is analyzed by Barbe and Bertail [1995] and Gleason [1988].

$BC_a$ intervals and a computationally rapid approximation known as the ABC method are described in Efron and Tibshirani [1994, Chapter 14]. See, also, Tibshirani [1988]. For deriving improved two-sided confidence intervals, see the discussion by Loh following Hall[1988; pp. 972-976]. Bootstrap iteration is introduced in Hall [1986] and Hall and Martin [1988].

Potential flaws in the bootstrap approach are considered by Schenker [1985], Efron [1988, 1992], Knight [1989], and Gine and Zinn [1989].

## 5.7   Exercises

1. Use the binomial formula of Section 4.2 to derive the values of $p$ depicted in Figure 5.2. Here is a time-saving hint: the probability of $k$ or more successes in $n$ trials is the same as the probability of no more than $n - k$ failures.

2. Which sex has longer teeth? Here are the mandible lengths of nine golden jackals (in mm):
   Males 107 110 116 114 113;
   Females 110 111 111 108.
   Obtain 90% confidence interval for the difference in tooth length. Apply both the bootstrap and the permutation test and compare the intervals you derive.

3. Exercise 7 of Chapter 3 described the results of an experiment with an antibiotic used to treat Herpes virus type II. Using the logarithm of the viral titer, determine an approximate 90% confidence interval for the treatment effect. (Hint: Keep subtracting a constant from the logarithms of the observations on saline controls until you can no longer detect a treatment difference.) Compare the intervals you derive using the bootstrap and the permutation test.

4. In Table 3.1, breaks appear to be related to log dose.

   **a.** Estimate $\rho$.

   **b.** Use both permutations and bootstrap to derive confidence intervals for $\rho$.

5. What is the correlation between undergraduate GPA and the LSAT? Use the law school data in Exercise 1.14 and the bootstrap to derive a confidence interval for the correlation between them.

6. Suppose you wanted to estimate the population variance. Most people would recommend you divide the sum of squares about the sample mean by the sample size minus 1, while others would say, too much work, just use the plug-in estimate. What do you suppose the difference is? (Need help deciding? Try this experiment: Take successive bootstrap samples of size 11 from the sixth-grade height data in Chapter 1 and use them to estimate the variance of the entire class.)

7. "Use the sample median to estimate the population median and the sample mean to estimate the population mean." This sounds like good advice, but is it? Use the technique described in Exercise 6 to check it out.

8. **a.** Obtain bootstrap confidence intervals for the variances of the billing values for Hospitals 1 and 4 (Exercise 1.16).

   **b.** Test the hypothesis that the population variance of Hospital 1 is equal to the sample variance of Hospital 4.

   **c.** Test the hypothesis that the population variances are the same for the two hospitals.

   **d.** Test the hypothesis that the population means are the same for the two hospitals.

**9.** Obtain smoothed bootstrap confidence intervals for
   **a.** the median of the student height data,
   **b.** correlation coefficient for the law school data.

**10.** Find a 98% confidence interval for the difference in test scores observed in Exercise 12 of Chapter 3. Does this interval include zero? If the new teaching method does not represent a substantial improvement, would you expect the confidence interval to include zero?

**11.** Should your tax money be used to fund public television? When a random sample of adults were asked for their responses on a 9-point scale (1 is very favorable and 9 is totally opposed) the results were as follows:
   3, 4, 6, 2, 1, 1, 5, 7, 4, 3, 8, 7, 6, 9, 5
   Provide a point estimate and confidence interval for the mean response.

**12.** Over 200 widgets per hour roll off the assembly line to be packed 60 to a carton. Each day you set aside 10 of these cartons to be opened and inspected. Today's results were 3, 2, 3, 1, 2, 1, 2, 1, 3, and 5 defectives.
   **a.** Estimate the mean number of defectives per carton and provide an 80% confidence interval for the mean.
   **b.** Can you be 90% certain the proportion of defectives is less than 5%?
   **c.** How does this exercise differ from Problem 1? (Hint: See Section 4.3.)

**13.** Apply the bootstrap to the billing data provided in Exercise 16 of Chapter 1 to estimate the standard error of the mean billing for each of Hospitals 1 and 2. Compare the ratio of the two standard errors with the ratio of the standard deviations. Why are these ratios so different? Check your intuition by computing the standard error and standard deviation for Hospital 3.

**14.** Can noise affect performance on the job? The following table summarizes the results of a time and motion study conducted in your factory.

| Noise (db) | Duration (min) | Noise (db) | Duration (min) | Noise (db) | Duration (min) |
|---|---|---|---|---|---|
| 0 | 11.5 | 40 | 15.0 | 80 | 28.5 |
| 0 | 9.5 | 40 | 16.0 | 80 | 27.5 |
| 20 | 12.5 | 60 | 20.0 | | |
| 20 | 13.0 | 50 | 19.5 | | |

Draw a graph of duration as a function of noise. Use Pitman correlation to determine if the relationship is significant. Use the bootstrap to obtain a confidence interval for the correlation.

# CHAPTER 6

# Power of a Test

In Chapter 3, you were introduced to some practical, easily computed tests of hypotheses. But are they the best tests one can use? And are they always appropriate? In this chapter, we consider the assumptions that underlie a statistical test and look at some of a test's formal properties: its significance level, power, and robustness.

In our example of the missing labels in Chapter 3, we introduced a statistical test based on the random assignment of labels to treatments, a permutation test. We showed this test provided a significance level of 5%, an *exact* significance level, not an approximation. The test we derived is valid under very broad assumptions. The data could have been drawn from a normal distribution (see Figure 4.2) or they could have come from some quite different distribution like the exponential (Figure 4.1). All that is required for our permutation test to be valid is that under the null hypothesis the distribution from which the data in the treatment group are drawn be the same as those from which the untreated sample is taken.

This freedom from reliance on numerous assumptions is a big plus. The fewer the assumptions, the fewer the limitations, and the broader the potential applications of a test. But before statisticians introduce a test into their practice, they need to know a few more things about it:

How powerful a test is it? That is, how likely is it to pick up actual differences between treated and untreated populations? Is this test as or more powerful than the test they are using currently?

How robust is the new test? That is, how sensitive is it to violations of the underlying assumptions and conditions of an experiment?

What if data are missing as is the case in so many practical experiments we perform? Will missing data affect the significance level?

What are the effects of extreme values or *outliers*? Can we extend the method to other, more complex experimental designs in which there are several treatments at several different levels and several simultaneous observations on each subject?

The balance of this chapter provides a theoretical basis for the answers.

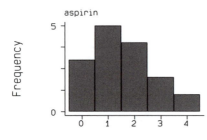
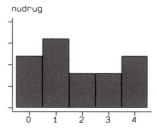

**Figure 6.1.** Patient self-rating in response to treatment.

Table 6.1. Decision Making Under Uncertainty

| The Facts | | Our Decisions |
|---|---|---|
| No Difference | No Difference | Drug is better Type I error: Manufacturer misses opportunity for profit. Public denied access to effective treatment. |
| Drug is better | Type II error: Manufacturer wastes money developing ineffective drug. | |

## 6.1   Fundamental Concepts

In this section, you are introduced in an informal way to the fundamental concepts of Type I and Type II error, significance level, power, and exact and unbiased tests.

### 6.1.1   Two Types of Error

Figure 6.1 depicts the results of an experiment in which two groups were each given a "painkiller." The first group got buffered aspirin, the second group received a new experimental drug. Each participant provided a subjective rating of the effects of the drug. The ratings ranged from "got worse," to "much improved," depicted below on a scale of 0 to 5. Take a close look at this figure. Does the new drug represent an improvement over aspirin?

Some of those who took the new experimental drug do seem to have done better. But not everyone. Are the differences we observe in Figure 6.1 simply the result of chance? Or do they represent a true treatment effect? If it's just a chance effect and we opt in favor of the new drug, we've made an error. We also make an error if we decide there is no difference and the new drug really is better. These decisions and the effects of making them are summarized in Table 6.1.

We distinguish the two types of error because they have the quite different implications described Table 6.1. As a second example, Fears, Tarone, and Chu

Table 6.2. Decision Making Under Uncertainty

| The Facts | Fears et al's Decisions | |
|---|---|---|
| No effect | Not a Carcinogen | Compound a Carcinogen Type I error: Manufacturer misses opportunity for profit. Public denied access to effective treatment. |
| Carcinogen | Type II error: Patients die; family suffer. Manufacturer sued. | |

[1977] use permutation methods to assess several standard screens for carcino-genicity. As shown in Table 6.2 their Type I error, a false positive, consists of labeling a relatively innocuous compound as carcinogenic. Such an action means economic loss for the manufacturer and the denial of the compound's benefits to the public. Neither consequence is desirable. But a false negative, a Type II error, would mean exposing a large number of people to a potentially lethal compound.

Because variation is inherent in nature, we are bound to make errors when we draw inferences from experiments and surveys, particularly if chance hands us a completely unrepresentative sample. When I toss a coin in the air six times, I can get three heads and three tails, but I can also get six heads. This latter event is less probable, but it is not impossible. Does the best team always win?

*We can't eliminate risk in decision making, but we can contain it* by the correct choice of statistical procedure. For example, we can require the probability of making a Type I error not exceed 5% (or 1% or 10%) and restrict our choice to statistical methods that ensure we do not exceed this level. If we have a choice of several statistical procedures all of which restrict the Type I error appropriately, we can choose the method which leads to the smallest probability of making a Type II error.

## 6.1.2  Losses

The preceding discussion is greatly oversimplified. Obviously, our losses will de-pend not merely on whether we guess right or wrong, but on how far our guessti-mate is off the mark. For example, you've developed a new drug to relieve anxiety and are investigating its side effects. Does it raise blood pressure? You do a study and find the answer is no. Alas, the truth is your drug raises systolic blood pres-sure an average of 1 mm. What is the cost to the average patient? Not much, negligible.

Now, suppose your new drug actually raises blood pressure an average of 10 mm. What is the cost to the average patient? to the entire potential patient population? to your company in lawsuits? One thing is sure: the cost of making a Type II error will depend on the magnitude of that error.

## 6.1.3  Significance Level and Power

In selecting a statistical method, statisticians work with two closely related concepts, significance level and power. The *significance level* of a test, denoted throughout the text by the Greek letter $\alpha$ (alpha), is the probability of making a Type I error; that is, $\alpha$ is the probability of deciding erroneously on the alternative when the hypothesis is true.

To test a hypothesis, we divide the set of possible outcomes into two or more regions. We reject the hypothesis and risk a Type I error when our test statistic lies in the *rejection region R*; we accept the hypothesis and risk a Type II error when our test statistic lies in the *acceptance region A*; and we may take additional observations when our test statistic lies in the boundary region of *indifference, I*. If $H$ denotes the hypothesis, then

$$\alpha = \Pr\{S \in A | H\}.$$

The *power* of a test, denoted throughout the text by the Greek letter $\beta$ (beta), is the complement of the probability of making a Type II error; that is, $\beta$ is the probability of deciding on the alternative when the alternative is the correct choice. If $K$ denotes the alternative, then

$$\beta = \Pr\{S \in R | K\}.$$

The ideal statistical test would have a significance level $\alpha$ of zero or 0% and a power $\beta$ of 1 or 100%. But unless we are all-knowing, this ideal cannot be realized. In practice, we fix a significance level $\alpha > 0$, where $\alpha$ is the largest value we feel comfortable with, and choose a statistic that maximizes or comes closest to maximizing the power.

### Power and Sample Size

As we saw in Section 6.1.2, the greater the discrepancy between the true alternative and our hypothesis, the greater the loss associated with a Type II error. Fortunately, in most practical situations, we can devise a test where the larger the discrepancy, the greater the power and the less likely we are to make a Type II error.

Figure 6.2 depicts the power as a function of the alternative for two tests based on samples of size 6. In the example illustrated, the test $\phi_1$ is uniformly more powerful than $\phi_2$, so that using $\phi_1$ in preference to $\phi_2$ will expose us to less risk.

Figure 6.3 depicts the power curve of these same two tests but using different size samples; the power curve of $\phi_1$ is still based on a sample of size 6, but that of $\phi_2$ now is based on a sample of size 12. The two new power curves almost coincide, revealing the two tests now have equal risks. But we will have to pay for twice as many observations if we use the second test in place of the first.

Moral: *A more powerful test reduces the costs of experimentation along with minimizing the risk.*[1]

---

[1]The exception proves the rule. When data gathering is dirt cheap, Lloyd Nelson observes, a less powerful test such as the sign test makes economic sense.

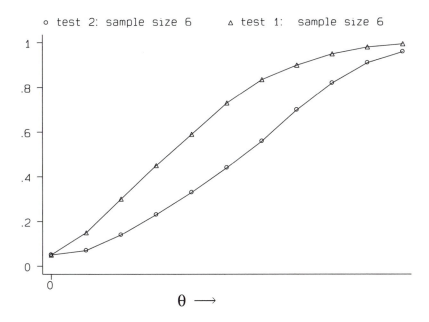

**Figure 6.2.** Power as function of the alternative; tests have same sample size.

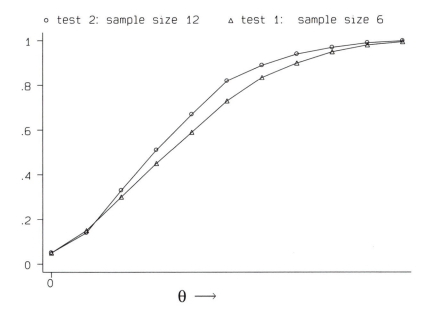

**Figure 6.3.** Power as function of then alternative; tests have different sample sizes and confidence intervals.

## Confidence Intervals and Sample Size

In Section 5.2, we saw the close relation between confidence intervals and tests of hypotheses. As the sample size increases, the width of our confidence interval decreases for a given level of significance and is less likely to include misleading values of the parameter we are estimating. If we keep the width of our confidence level fixed, then as the sample size increases, we can have increasingly greater confidence that it includes the correct parameter value.

---

*Power* of a test depends upon the statistic, the sample size, and the alternative.

*Width* of a confidence interval depends upon the statistic, the sample size, and the confidence level.

---

## Power and the Alternative

If a test at a specific significance level $\alpha$ is more powerful against a specific alternative than all other tests at the same significance level, we term it *most powerful*. But as we see in Figure 6.4, a test that is most powerful for some alternatives may be less powerful for others. For near alternatives, with $\theta$ close to zero, test 4 is the more powerful test; for far alternatives with $\theta$ large, test 3 is more powerful. Thus neither test is uniformly most powerful. When a test at a specific significance level is more powerful against all alternatives than all other tests at the same significance level, we term this test *uniformly* most powerful.

The significance level and power may also depend upon how the variables we observe are distributed. Does the population distribution follow a bell-shaped normal curve with the most frequent values in the center? or is the distribution something quite different? To protect our interests, we may need to require that the Type I error be less than or equal to some predetermined value for all possible distributions. When applied correctly, with the assumption of exchangeability satisfied, permutation tests always have this property. The significance level of a test based on the bootstrap is dependent on the underlying distribution.

## 6.1.4   Exact, Unbiased Tests

In practice, we seldom know the distribution of a variable or its variance. Suppose we wish to test the hypothesis "$X$ has mean 0." *This compound hypothesis* includes several *simple hypotheses* such as $H_1$: $X$ is normal with mean 0 and variance 1, $H_2$: $X$ is normal with mean 0 and variance 1.2, and $H_3$: $X$ is a gamma distribution with mean 0 and four degrees of freedom.

A test is said to be *exact* with respect to a compound hypothesis if the probability of making a Type I error is exactly $\alpha$ for each and every one of the possibilities that make up the hypothesis. A test is said to be *conservative* if the Type I error never exceeds $\alpha$. Obviously, an exact test is conservative though the reverse may not be true.

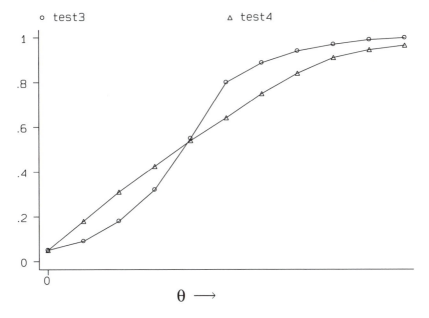

**Figure 6.4.**  Comparing power curves.

The importance of an exact test cannot be overestimated, particularly a test that is exact regardless of the underlying distribution. If a test that is nominally at level $\alpha$ is actually at level $\gamma$, we may be in trouble before we start: If $\gamma > \alpha$, the risk of a Type I error is greater than we are willing to bear. If $\gamma < \alpha$, then our test is suboptimal, and we can improve on it by enlarging its rejection region. We return to these points again in Chapter 11 on choosing a statistical method.

A test is said to be *unbiased* and of level $\alpha$ providing its power function $\beta$ satisfies the following two conditions:

1. $\beta$ is *conservative*; that is, $\beta(\theta) \leq \alpha$ for every $\theta$ that satisfies the hypothesis,

2. $\beta(\theta) \geq \alpha$ for every $\theta$ that is an alternative to the hypothesis.

In other words, a test is unbiased if it is more likely to reject a false hypothesis than a true one. If $\{A(\theta)\}$ is a collection of acceptance regions associated with a uniformly most powerful unbiased test, the correct value of the parameter is more likely to be covered by the confidence intervals we construct from $\{A(\theta)\}$ than an incorrect value.

Faced with some new experimental situation, our objective always is to derive a uniformly most powerful unbiased test if one exists. But, if we can't derive a uniformly most powerful test, and Figure 6.4 depicts just such a situation, then we will look for a test which is most powerful against those alternatives that are of immediate interest.

## 6.1.5    Dollars and Decisions

We now know a statistical problem is defined by four elements:

1. the observations $X = \{X_1, X_2, \ldots, X_n\}$,

2. the distribution $F$ of the variables $\{X_i\}$ in the population from which the observations are drawn,

3. the set $\{d\}$ of possible decisions one can make on observing $X$,

4. the loss $L(d, \theta)$, expressed in dollars, lives, or some other measure, that results when we make the decision $d$ when $\theta$ is true.

A problem is a statistical one when the investigator is not in a position to say that $X$ will take on exactly the value $x$, but only that $X$ has some probability $P\{x \in A\}$ of taking on values in the set $A$.

In this text, we've limited ourselves for the most part to two-sided decisions in which either we accept a hypothesis $H$ and reject an alternative $K$, or we reject the hypothesis $H$ and accept the alternative $K$. One example is the hypothesis that the population mean is less than or equal to $\theta_0$ versus the alternative that it is greater. As in Section 5.2, we would probably follow our decision to accept or reject with a confidence interval for the unknown parameter, such as $\theta_1 <$ Mean $< \theta_2$, and a statement to the effect that the probability this interval covers the true parameter value is not less than $1 - \alpha$.

Typically, losses $L$ depend on some function of the difference between the true (but unknown) value of the parameter we are trying to estimate $\theta$ and our best guess (point estimate) of its value $\theta^*$; $L(\theta, \theta^*) = |\theta - \theta^*|$, for example, or $L(\theta, \theta^*) = (\theta - \theta^*)^2$. Our objective is to come up with a decision rule $D$, such that when we average over all possible samples we might draw from the population F, we minimize the associated *risk*

$$R_\theta(D) = \sum L(D[x_1, x_2, \ldots, x_n]) \Pr\{x_1, x_2, \ldots, x_n | \theta\}.$$

Unfortunately, a testing procedure that is optimal for one value of the parameter $\theta$ might not be optimal for another. This situation is illustrated in Figure 6.4 with two decision curves that cross each other. My tentative solution to this complex problem is to always use an unbiased procedure.[2]

## 6.1.6    What Significance Level Should I Use?

Choice of significance level and power are determined by the environment in which you work.

Most scientists simply report the observed $p$-value leaving it to their peers to decide for themselves what weight should be given the results.

---

[2]This problem is complex with philosophical as well as mathematical overtones; we refer the interested reader to the discussions in the first chapter of Erich Lehmann's book on *Testing Statistical Hypothesis* [1986]. A dissenting view will be found in Suissa and Shuster [1984].

---

### Significance Level, Power, and Sample Size

Significance level, power, and sample size are interrelated. Fix the significance level at a given sample size and you determine the power against all alternatives. Increase the sample size without changing the significance level and you increase the power. If you are willing to accept a greater risk of making a Type I error, then you can take fewer observations without affecting the power. Of course, both power and significance level depend upon your choice of test statistic.

---

A manufacturer preparing to launch a new product line or a pharmaceutical company conducting a research for promising compounds typically adopt a three-way decision procedure: If the observed $p$-value is less than 1%, they go forward with the project. If the $p$-value is greater than 20%, they abandon it. And if the $p$-value lies in the gray area in between, they arrange for additional surveys or experiments.

A regulatory commission like the FDA that is charged with oversight responsibility must work at a fixed significance level, typically $\alpha = 0.05$ or $0.10$. The choice of a fixed significance level ensures consistency in both result and interpretation as the agency reviews the findings from literally thousands of tests. They specify a minimum power level of 90% or 95%, or even 99% against biologically significant alternatives if potentially dangerous side effects are under investigation.

## 6.2  Assumptions

Every test carries with it certain assumptions, that the sample was selected at random from the population, for example. These assumptions must be satisfied in order to achieve a desired significance level.

Many of the statistical procedures in common use (see Chapter 4) rely on the distribution of the underlying data having a specific form, such as the normal, or binomial, or Poisson. If this assumption is correct, *parametric* statistical procedures such as the $t$ test or the $F$ test may be called for. If the assumption is in error, parametric procedures may yield erroneous results.

Underlying the bootstrap are two assumptions: a) if the null hypothesis is true, then all observations in the sample(s) come from populations that have the same value of the parameter that is being estimated and b) the observations are independent of one another.

Underlying all permutation tests is the assumption that, if the null hypothesis is true, all observations in the sample (or subsample) are *exchangeable*. Loosely speaking, observations are exchangeable if either they are

    a. independent and identically distributed,

    b. identically normally distributed with the same mutual correlations,

c. obtained by sampling without replacement from the same finite population.

Sometimes a simple transformation will ensure observations are exchangeable or, in the case of the bootstrap, have the same parameter. For example, if we know that $X$ comes from a population with mean $\mu$ and distribution $F[x - \mu]$ and an independent observation $Y$ comes from a population with mean $v$ and distribution $F[x - v]$, then the independent variables $X' = X - \mu$ and $Y' = Y - v$ are exchangeable.

## 6.2.1   Transformations

We may need to transform our data before we analyze it for any or all of the following reasons:

1. to permit us to test a *null* hypothesis as in the preceding example,
2. to obtain a more symmetric distribution,
3. to equalize the variances (and, hopefully, the distributions),
4. to reduce the impact of outliers.

Recall that more accurate results can be obtained if the bootstrap distribution is symmetric and that to perform a one-sample permutation test the observations must come from a symmetric distribution. Taking logarithms of the observations will often produce a more symmetric distribution. If we anticipate changes measured in percentage rather than absolute terms, a preliminary logarithmic transformation should always be used.

The key issue with permutation tests is whether, under the null hypothesis of no differences among the various experimental or survey groups, we can exchange the labels on the observations without affecting the results. Suppose we have taken steps to ensure our sampling method is representative (as described in Section 2.6), and that our observations are independent of one another. Certain equalizing transformations then can help us to ensure the variances (and, hopefully, the distributions) of the variables we are comparing are the same.

If the variances are expected to be proportional to the mean, take the logarithms of the observations. If the standard deviations are expected to be proportional to the mean as is the case with Poisson variables, take the square roots of the observations. With proportions such as the proportion of binomial trials that result in success, use the arcsin transformation.

Not infrequently, we encounter values among our observations that seem to stand out from the rest, a "19" for example in Section 3.2 when all the rest of the observations lay between 1 and 6. Is the "19" a typographical error? Should it be 1.9 instead, or simply 1, or 9? When we can't be sure, replacing the observations by their ranks will at least reduce the impact of exceptional values, while not reducing their effect completely in case they're real. Replacing observations by their ranks does not affect the significance level if done routinely and has the same effect with large samples on the power of a two-sample comparison as reducing the sample size from 100 to between 86 and 96 (Bradley, 1968).

---

**Which Test?**

If you can exchange the labels on the observations without affecting the results, it is safe and preferable to use a permutation test.

If the observations have the same population parameters when the null hypothesis is true, it is safe to use the bootstrap.

---

## 6.3    How Powerful Are Our Tests?

Given that these assumptions are satisfied, how effective and how powerful are the tests we introduced in preceding chapters?

### 6.3.1    One Sample

Our permutation test is applicable even if the different observations come from different distributions providing that under the null hypothesis these distributions are all symmetric and all have the same median.[3] Against specific normal alternatives, our permutation test provides a most powerful unbiased test of the distribution-free hypothesis $H$: $\theta = \theta_0$.[4] Even if we know the underlying data have the normal distribution (and it is rare we would know), for large samples, the permutation test's power is almost the same as the most powerful parametric test, Student's $t$.[5] Even if the underlying distributions are almost but not quite symmetric, Romano [1990] shows that for very large samples, the one-sample permutation test for a location parameter is exact providing the underlying distribution has finite variance. His result applies whether the permutation test is based on the mean, the median, or some statistical functional of the location parameter. If the underlying distribution is almost symmetric, the permutation test will be almost exact even when based on as few as 10 or 12 observations.

The bootstrap is neither exact nor conservative, though if the observations are independent and from distributions with identical values of the parameter of interest, the bootstrap is exact for very large samples of tens of thousands of observations.[6] A nonparametric bootstrap does not provide exact significance levels and is less powerful than a permutation test, but in the one-sample case does not require the underlying distribution to be symmetric.[7] The observations can even come from different distributions providing all have the same location parameter.

---

[3] A simple transformation can often ensure this requirement is satisfied; for example, subtracting a constant from all members of the second sample, or working with logarithms rather than the original observations. See Appendix 1.

[4] Lehmann [1986, p. 239].

[5] Albers, Bickel, and van Zwet [1976].

[6] Liu [1988].

[7] As noted in Chapter 4, when the underlying distribution is not symmetric, a sample with at least 50 observations is essential if the bootstrap is to be a valid approximation.

---

**Guidelines for Bootstrap Hypothesis Testing**

The bootstrap is neither exact nor conservative and may have very low power even against distant alternatives. To increase power and improve accuracy, Hall and Wilson [1991] provide the following guidelines for a test of $H: \theta = \theta_0$:

1. Resample $\widehat{\theta}^* - \widehat{\theta}$ rather than $\widehat{\theta}^* - \theta_0$, where $\widehat{\theta}$ is the estimate of $\theta$ obtained from the original sample, and $\widehat{\theta}^*$ is the estimate obtained from the bootstrap sample.

2. Base the test on the bootstrap distribution of $(\widehat{\theta}^* - \widehat{\theta})/\widehat{\sigma}^*$, where $\widehat{\sigma}^*$ is an estimate of the standard deviation of $\widehat{\theta}$ based upon the bootstrap sample, rather than on the bootstrap distribution of $(\widehat{\theta}^* - \widehat{\theta})/\widehat{\sigma}$ or $\widehat{\theta}^* - \widehat{\theta}$.

---

## 6.3.2   Matched Pairs

The results for a single sample apply here as well, with the added bonus that we know the underlying data is symmetric.[8]

## 6.3.3   Two Samples

A permutation test based on our statistic is exact and unbiased against stochastically increasing shift alternatives of the form $K: F_1[x] = F[x - d]$, for $d > 0$. In fact, this permutation test is a uniformly most powerful unbiased test of the null hypothesis $H: F_1 = F$, against normally distributed shift alternatives. Against normal alternatives and for samples of 20 or more, its power is equal to that of the standard $t$ test described in Section 4.3.2.[9]

The permutation test offers the advantage over the parametric $t$ test that it is exact even for very small samples *whether or not the observations come from a normal distribution*. The parametric $t$ test relies on the existence of a mythical infinite population from which all the observations are drawn. The permutation test is applicable even to finite populations such as all the machines in a given shop or all the supercomputers in the world.

The bootstrap does not require the observations all come from the same population, only that they all have the same value of the population parameter under investigation. The bootstrap is neither exact nor conservative, though if the observations are independent and from distributions with identical values of the parameter of interest, then the bootstrap is exact for very large samples [Liu, 1988]. The number of observations required for an almost-exact bootstrap can be reduced by using $BC_a$ intervals as described in Section 5.4 and the Hall-Wilson guidelines outlined in the accompanying sidebar..

A nonparametric bootstrap is less powerful than a permutation test. Still, the bootstrap may be applicable when neither a permutation test nor a parametric test exists.

---

[8] See, for example, Good [2000, p. 52].
[9] Bickel and van Zwet [1978].

## 6.4 Which Test?

We are now able to make an initial comparison of the three types of statistical tests—permutation, bootstrap, and parametric.

Recall from Chapter 3 that with a permutation test, we

1. Choose a test statistic $S(X)$.

2. Compute $S$ for the original set of observations.[10]

3. Obtain the permutation distribution of $S$ by repeatedly rearranging the observations. With two or more samples, we combine all the observations into a single large sample before we rearrange them.

4. For a one-sided test, obtain the upper $\alpha$-percentage point of the permutation distribution and accept or reject the null hypothesis according to whether $S$ for the original observations is smaller or larger than this value.

If the labels on the observations are exchangeable, then the resultant test is exact and unbiased.

The nonparametric bootstrap, like the permutation test, requires a minimum number of assumptions and derives its critical values from the data at hand.

Recall from Chapter 5 that to obtain a nonparametric bootstrap, we

1. Choose a test statistic $S(X)$.

2. Compute $S$ for the original set of observations.

3. Obtain the bootstrap distribution of $S$ by repeatedly resampling with replacement from the observations in a way that is consistent with the null hypothesis. We need not combine the samples, but may resample separately from each. We can take advantage of any additional information such as symmetry or (via a parametric bootstrap) any more detailed knowledge of the distribution.

4. Smooth the bootstrap distribution and/or apply bias corrections.

5. For a one-sided test, obtain the upper $\alpha$-percentage point of the corrected bootstrap distribution and accept or reject the null hypothesis according to whether $S$ for the original observations is smaller or larger than this value.

Recall from Chapter 4 that to obtain a parametric test (e.g., a $t$ test or an $F$ test), we

1. Choose a test statistic $S$ whose distribution $F_S$ may be computed and tabulated independent of the observations.

2. Compute $S$ for the observations $X$.

3. For a one-sided test, compare $S(X)$ with the upper $\alpha$-percentage point of $F_S$ and accept or reject the null hypothesis according to whether $S(X)$ is smaller or larger than this value.

---

[10]We may need to transform the observations first.

If the observations are independent, the sample selected by representative means, and $S$ is distributed as $F_S$, then the parametric test is exact and, often, the most powerful test available. If $S$ really has some other distribution, then the parametric test may lack power and may not be conservative. With large samples, the permutation test is usually as powerful as the most powerful parametric test.[11] If $S$ is not distributed as $F_S$, it may be more powerful. How large is large? If we are investigating a location parameter, it could be as few as 10 or 12 observations. In other instances, involving ratios of several parameters, large could mean 100 or more.

## 6.5    Summary

In this chapter, you learned that power, sample size, and significance level are interrelated. You learned the value of exact unbiased tests. You learned that your choice of test statistic will depend on the hypothesis, the alternative, the loss function, and the type of test you employ. You learned the assumptions underlying and the differences among test of hypotheses based on the bootstrap, the permutation test, and parametric distributions.

## 6.6    To Learn More

A formal discussion of risk theory is found in Lehmann [1986]. For a further discussion of exchangeability, see Lehmann [1986, p. 231], Koch [1982], and Draper et al. [1993]. Examples of power calculations for resampling methods are given in Oden [1975], Keller-McNulty and Higgens [1987], and Hall and Titterington [1989]. For a discussion of the relative advantages of variable versus fixed significance levels, see Kempthorne [1966].

Singh [1981] was the first to demonstrate the advantages of the bootstrap over large-sample parametric approximations. For other large-sample results, see Good [2000].

## 6.7    Exercises

1. **a.** Sketch the power curve $\beta(\theta)$ for one of the two-sample comparisons described in Chapter 3. (You already know two of the values for each power curve. What are they?)

   **b.** Using the same set of axes, sketch the power curve of a test based on a much larger sample.

---

[11] Bickel and Van Zwet [1978].

**c.** Suppose that without looking at the data you i) always reject; ii) always accept; or iii) use a chance device so as to reject with probability $\alpha$.

For each of these three tests, determine the power and the significance level. Are any of these three tests exact? unbiased?

2. True or False? Tests should be designed so that
   **a.** the risk of making a Type I error is low.
   **b.** the probability of making a Type II error is high.
   **c.** the null hypothesis is likely to be rejected.
   **d.** economic consequences of a decision are considered.

3. True or False? The expected loss from using a particular statistical procedure will depend on
   **a.** the probability the procedure will lead to a wrong decision.
   **b.** the losses associated with a wrong decision.

4. **a.** Suppose you have two potentially different radioactive isotopes with half-life parameters $\lambda_1$ and $\lambda_2$, respectively. You gather data on the two isotopes and, taking advantage of a uniformly most powerful unbiased permutation test, you reject the null hypothesis $H: \lambda_1 = \lambda_2$ in favor of the one-sided alternative $K: \lambda_1 > \lambda_2$. What are you or the person you are advising going to do about it? Will you need an estimate of $\lambda_1/\lambda_2$? What estimate will you use?

   **b.** Review some of the hypotheses you tested in the past. Distinguish your actions after the test was performed from the conclusions you reached. (In other words, did you do more testing? rush to publication? abandon a promising line of research?) What losses were connected with your actions? Should you have used a higher/lower significance level? Should you have used a more powerful test or taken more or fewer observations? And, if you used a parametric test like Student's $t$ or Welch's $z$, were all the assumptions for these tests satisfied?

5. Late for a date, you dash out of the house, slipping a bill into your wallet. Was it a $2 bill or a $20? Too late now. At the movie theater, your friend slips a $2 bill into your hand, "for my share." (A share of what? the popcorn?) You put this bill too away in your wallet. As you approach the ticket window, you decide to perform a simple statistical test. You'll take a quick look at one of the bills. If it's a $2 bill, you'll accept the hypothesis that both bills are $2. Otherwise, you'll reject this hypothesis. What is the significance level of your test? the power?

6. Given a choice of a permutation test based on the original scores, a permutation test based on ranks, a bootstrap, and the best and most appropriate parametric test, which would you use in the following situations and why?

[Hint: You may want to glance through Chapters 3, 4, and 5 a second time.]
In cases a–e, you want to test a hypothesis about the median of a population:

**a.** You have 10 observations from a normal distribution.

**b.** You have 10 observations from an almost-normal distribution.

**c.** You have 10 observations from an exponential distribution.

**d.** You have 10 observations from an almost-exponential distribution.

**e.** You have 50 observations from an almost-exponential distribution.

In cases f–h, you want to compare two populations.

**f.** You have two samples of size 10 taken from normal distributions with the same variance and want to compare medians.

**g.** You have two samples of size 10 taken from similarly shaped distributions and want to compare medians. One problem: all the data are in the hundreds, except for one observation which is 10.1. Or is that decimal point a flaw in the paper?

**h.** You have two samples of size 25 and want to compare variances.

7. **a.** Your lab has been gifted with a new instrument offering 10 times the precision of your present model. How might this affect the power of your tests? their significance level? the number of samples you'll need to take?

**b.** A directive from above has loosened the purse strings so you now can take larger samples. How might this affect the power of your tests? their significance level? the precision of your observations? the precision of your results?

**c.** A series of lawsuits over those silicon implants you thought were harmless has totally changed your company's point of view. How might this affect the power of your tests? their significance level? the precision of your observations? the number of samples you'll take? the precision of your results?

8. Turn to the literature of your field and consider some of the estimates that have been made. Which do you feel is the most appropriate loss function in each case $L_1 = |\theta - \theta^*|$ or $L_2 = (\theta - \theta^*)^2$? If hypotheses were tested, were all test assumptions satisfied?

9. Imagine your company, a manufacturer of industrial supplies, plans to launch a new product to sell to existing customers. Development costs would be approximately one million dollars. Your anticipated profit is $1000 per unit sold. You have a customer base of 10,000. What percentage would you need to sell to break even? Of course the actual percentage you sell might be greater or less than this amount. If you decide to manufacture the product, what sort of losses might you experience? If you decide against development, what profits might you be surrendering?

You decide to do a survey of your customers and use this survey to estimate the percentage you will sell. If $p$ is your estimate and $\pi$ the actual value of this percentage, construct a graph of losses in dollars versus the estimate error $(\pi - p)$.

10. Your company policy is to insist on a 99% compliance policy in the components you purchase. That is, no more than 1% of the units can be out of compliance. How many units would you need to examine to be sure of rejecting a lot shipped to you with 2% defective components? Examining all of them would get the job done, but there must be a less expensive way. Make a chart for $n = 100, 200$, and so forth, in which you list $k$ (the maximum number of tested units that can be out of compliance), $\alpha$ (the resulting significance level when 99% of the components are actually in compliance), and $\beta$ (the power when only 98% of the units are actually in compliance).

11. Take a second look at the experimental methods described in Exercises 11 and 12 of Chapter 3. Which approach ought to yield the more powerful test?

12. Take a second look at the process temperature data of Cox and Snell [1981] in Table 3.2. What would be the effect on the significance level and power of a two-tailed bootstrap test of the hypothesis $\theta = 440$ if we follow the first of the Hall–Wilson guidelines? the second guideline?

13. Formulate hypotheses and alternatives for comparing the billing practices of the four hospitals considered in Exercise 16 of Chapter 1. What will you compare? means? medians? variances? frequency distributions? (You may wish to return to this question after completing Chapters 7 and 8.)

# CHAPTER 7

# Categorical Data

In many experiments and in almost all surveys, many if not all the results fall into categories rather than being measurable on a continuous or ordinal scale: male vs. female, black vs. Hispanic vs. oriental vs. white, in favor vs. against vs. undecided. The corresponding hypotheses concern proportions: "Blacks are as likely to be Democrats as they are to be Republicans." Or, "the dominant genotype 'spotted shell' occurs with three times the frequency of the recessive." In this chapter, you learn to test hypotheses like these that concern categorical and ordinal data.

## 7.1 Fisher's Exact Test

As an example, suppose on examining the cancer registry in a hospital, we uncover the following data which we put in the form of a $2 \times 2$ *contingency table*.

| 7.1 | Survived | Died | Total |
|-------|----------|------|-------|
| Men | 9 | 1 | 10 |
| Women | 4 | 10 | 14 |
| Total | 13 | 11 | 24 |

The 9 denotes the number of males who survived, the 1 denotes the number of males who died, and so forth. The four marginal totals or *marginals* are 10, 14, 13, and 11. The total number of men in the study is 10; 14 denotes the total number of women, and so forth.

We see in this table an apparent difference in the survival rates for men and women: Only 1 of 10 men died following treatment, but 10 of the 14 women failed to survive. Is this difference statistically significant?

The answer is yes. Let's see why, using the same line of reasoning that Fisher advanced at the annual Christmas meeting of the Royal Statistical Society in 1934. (After Fisher's talk was concluded, incidentally, a seconding speaker compared

Fisher's talk to "the braying of the Golden Ass." I hope you will take more kindly to my own explanation.) The preceding contingency table has several fixed elements:

- the total number of men in the survey, 10,
- the total number of women, 14,
- the total number who died, 11,
- and the total number who survived, 13.

These totals are immutable; no swapping of labels will alter the total number of individual men and women or bring back the dead. But these totals do not determine the contents of the table as can be seen from the two tables reproduced below whose marginals are identical with those of our original table.

| 7.2 | Survived | Died | Total |
|---|---|---|---|
| Men | 10 | 0 | 10 |
| Women | 3 | 11 | 14 |
| Total | 13 | 11 | 24 |

| 7.3 | Survived | Died | Total |
|---|---|---|---|
| Men | 8 | 2 | 10 |
| Women | 5 | 9 | 14 |
| Total | 13 | 11 | 24 |

The first of these tables makes a strong case for the superior fitness of the male, stronger even than our original observations. In the second table, the survival rates for men and women are more alike than they were in our original table.

Fisher would argue that if the survival rates were the same for both sexes, then each of the redistributions of labels to subjects, that is, each of the $N$ possible contingency tables with these same four fixed marginals, is equally likely, where,[1,2]

$$N = \sum_{x=0}^{10} \binom{13}{x}\binom{11}{10-x} = \binom{13+11}{10}.$$

How did we get this value for $N$? The component terms are taken from the hypergeometric distribution:

$$\sum_{x=0}^{t} \binom{m}{x}\binom{n}{t-x} \Big/ \binom{m+n}{t} \tag{7.1}$$

where $n$, $m$, $t$, and $x$ occur as the indicated elements in the following $2 \times 2$ contingency table

---

[1]Combinatorial notation, such as choose $x$ of 13 things, is defined in Section 2.5.1.

[2]You can illustrate this result by making up two different sets of labels. Take the first of these sets and mark the individual labels either "man" or "woman." Take the second set and mark each label as either "survived" or "died." Now give each member of your class one label selected at random from each set. If you're at home alone, you can hand out the labels to pillows or pieces of furniture.

|  | Category 1 | Category 2 |  |
|---|---|---|---|
| Category A | $x$ | $t - x$ | $t$ |
| Category B | $m - x$ | $n - (t - x)$ |  |
|  | $m$ | $n$ | $m + n$ |

If men and women have the same probability of surviving, then all tables with the marginals $m, n, t$ are equally likely, and $\sum_{k=0}^{t-x} \binom{m}{t-k}\binom{n}{k}$ are as or more extreme.

In our example, $m = 13$, $n = 11$, $x = 9$, and $t = 10$, so that $\binom{13}{9}\binom{11}{1}$ of the $N$ tables are as extreme as our original table and $\binom{13}{10}\binom{11}{0}$ are more extreme.

$11\binom{13}{9}+\binom{13}{10}$ is a very small fraction of the total, so we conclude that a difference in survival rates as extreme as the difference we observed in our original table is very unlikely to have occurred by chance. We reject the hypothesis that the survival rates for the two sexes are the same and accept the alternative that, in this instance at least, males are more likely to profit from treatment.

## 7.1.1    One-Tailed and Two-Tailed Tests

In the preceding example, we tested the hypothesis that survival rates do not depend on sex against the alternative that men diagnosed as having cancer are likely to live longer than women similarly diagnosed. We rejected the null hypothesis because only a small fraction of the possible tables were as extreme as the one we observed initially. This is an example of a one-tailed test. Or is it? Wouldn't we have been just as likely to reject the null hypothesis if we had observed a table of the following form:

| 7.4 | Survived | Died | Total |
|---|---|---|---|
| Men | 0 | 10 | 10 |
| Women | 13 | 1 | 14 |
| Total | 13 | 11 | 24 |

Of course, we would. In determining the significance level in the present example, we must add together the total number of tables which lie in either of the two extremes or tails of the permutation distribution.

McKinney et al. [1989] reviewed some 70 plus articles that appeared in six medical journals. In over half these articles, Fisher's exact test was applied improperly. Either a one-tailed test had been used when a two-tailed test was called for or the authors of the paper simply hadn't bothered to state which test they had used.

When you design an experiment, decide at the same time whether you wish to test your hypothesis against a two-sided or a one-sided alternative. A two-sided alternative dictates a two-tailed test; a one-sided alternative dictates a one-tailed test.

As an example, suppose we decide to do a follow-on study of the cancer registry to confirm our original finding that men diagnosed as having tumors live significantly longer than women similarly diagnosed. In this follow-on study, we have a one-sided alternative. Thus, we would analyze the results using a one-tailed test rather than the two-tailed test we applied in the original study.

---

**Warning**

McKinney et al. [1989] report that more than half the published articles that apply Fisher's exact test fail to make the correct distinction between one-tailed and two-tailed tests. Don't you make the same mistake!

---

## 7.1.2  The Two-Tailed Test

Unfortunately, it is not obvious which tables should be included in the second tail. Is Table 7.4 as extreme as Table 7.2? We need to define a test statistic to serve as a basis of comparison. One commonly used measure is the Pearson $\chi^2$ statistic defined for the $2 \times 2$ contingency table after eliminating elements that are invariant under permutations as $(x - t\frac{m}{m+n})^2$. This statistic is proportional to the square of the difference between what one would expect to observe if the survival rates are the same, that is, $t\frac{m}{m+n}$, and the frequency $x$ actually observed. For Table 7.1, this statistic is 13, for Table 7.4, it is 29. We leave it to you to do the computations to show that Table 7.5 is more extreme than 7.1, but 7.6 is not.

| 7.5 | Survived | Died | Total |
|---|---|---|---|
| Men | 1 | 9 | 10 |
| Women | 12 | 2 | 14 |
| Total | 13 | 11 | 24 |

| 7.6 | Survived | Died | Total |
|---|---|---|---|
| Men | 2 | 8 | 10 |
| Women | 11 | 3 | 14 |
| Total | 13 | 11 | 24 |

The Pearson statistic is far from being our only choice; among the others are the likelihood ratio statistic $x \log[xtm/(m+n)]$ and Fisher's statistic

$$-2 \log h[y] - \log[2.51(m+n)^{-3/2}\sqrt{mnt(m+n-t)}]$$

where $h[y]$ is the proportion of all tables with the same marginals that have precisely the same four entries. For very large samples with a large number of observations in each cell, all three statistics lead to the same conclusion.

## 7.2  Odds Ratio

In most instances, we won't be satisfied with merely rejecting the null hypothesis but will want to make a more powerful statement such as "men are twice as likely as women to get a good-paying job," or "women under thirty are twice as likely as men over 40 to receive an academic appointment."

For the survival data of Table 7.1, the odds ratio $\frac{\pi_2}{1-\pi_2}\big/\frac{\pi_1}{1-\pi_1}$ is 0.044 with a 90% confidence interval using the method of Gart [1971] extending from 0.002 to 0.44.

In the discrimination case of Fisher versus Transco Services of Milwaukee [1992], the plaintiffs claimed that Transco was 10 times as likely to fire older employees. Can we support this claim with statistics? The Transco data are provided in the following table.

Table 7.7. Transco Employment

| Outcome | Young | Old |
|---------|-------|-----|
| Fired | 1 | 10 |
| Retained | 24 | 17 |

Let $\pi_1$ denote the probability of firing a young person and $\pi_2$ the probability of firing an older person. We want to go beyond testing the null hypothesis $\pi_1 = \pi_2$ to determine a confidence interval for the odds ratio $\frac{\pi_2}{1-\pi_2}\big/\frac{\pi_1}{1-\pi_1}$. This problem is too complex, and there are too many rearrangements to rely on hand calculations. We turn for aid to StatXact™, a statistical package whose emphasis is the analysis of categorical and ordinal data. Choose Statistics, Two Binomials, and CI. Odds Ratio from successive StatXact menus. Based on the results depicted in the accompanying sidebar, we can tell the judge that older workers were fired at a rate at least 1.65 times the rate at which younger workers were discharged.

```
              ODDS RATIO OF TWO BINOMIAL PROPORTIONS

StatXact Output
Datafile: C:\EXAMPLES\TRNSCO.CY3

Statistic based on the observed 2 by 2 table:
    Binomial proportion for column <young>:pi_1 = 0.04000
    Binomial proportion for column <Old>: pi_2 = 0.3704
                  ( pi_2 )/(1-pi_2 )
    Odds Ratio = ------------------- = 14.12
                  ( pi_1 )/(1-pi_1 )
Results:
Method:               P-value(2-sided) 95.00% Confidence Interval
Asymp (Mantel-Haenszel)        0.0157  (1.649,120.9)
Exact                          0.007145 (1.649,637.5)
```

## 7.2.1    Stratified $2 \times 2$'s

In trying to develop a cure for a relatively rare disease, we face the problem of having to gather data from a multitude of test centers, each with its own set of procedures and its own way of executing them. Before we can combine the data, we must be sure the odds ratios across the test centers are approximately the same. Consider the set of results in Table 7.8 obtained by the Sandoz drug company and reproduced with permission from the StatXact manual. One of the cites, number 15, stands out from the rest. But is the difference statistically significant?[3] With 22 contingency tables, the number of computations needed to examine all rearrangements is in the billions.

Fortunately, StatXact utilizes several time-saving algorithms, including the one introduced in Mehta, Patel, and Senchaudhuri [1988], to obtain a Monte Carlo estimate of the significance level. We pull down menus Statistics, Stratified 2x2 Tables, and Homogeneity of Odds Ratios. The estimated $p$-value of 0.013, just a fraction greater than 1%, tells us it would be unwise to combine the results from the different cites.

The output of this program provides us with one more important finding: Displayed above the Monte Carlo estimate of the exact $p$-value, 0.01237, is the asymptotic or large-sample approximation based on the chi-square distribution.

Its value, 0.0785, is many times larger than the correct value, and relying on this so-called approximation would have led us to a completely different and erroneous conclusion.

---

**TEST FOR HOMOGENEITY OF ODDS RATIOS**
**[18 2x2 informative tables]**

```
StatXact Output
Datafile: C:\EXAMPLES\SANDOZ.CY3

Observed Statistics:
    BD: Breslow and Day Statistic = 25.78
    ZE: Zelen Statistic = 9.481e-009

Asymptotic p-value: (based on chi-square distribution with 17 df)
    Pr{BD.GE.25.78} = 0.0785

Monte Carlo estimate of p-value:
    Pr{ZE.GE.9.481e-009} = 0.0127
    99.00% Confidence Interval = (0.0119,0.0135)
```

---

[3]Similar problems were encountered in a study where test subjects might use one of several different "identical" machines. I couldn't combine the results from the different machines or the different technicians who operated them, until I'd performed an initial test of their equivalance.

Table 7.8. Sandoz Drug Data.

| Test site | New Drug | | Control Drug | |
|---|---|---|---|---|
| | Response | # | Response | # |
| 1 | 0 | 15 | 0 | 15 |
| 2 | 0 | 39 | 6 | 32 |
| 3 | 1 | 20 | 3 | 18 |
| 4 | 1 | 14 | 2 | 15 |
| 5 | 1 | 20 | 2 | 19 |
| 6 | 0 | 12 | 2 | 10 |
| 7 | 3 | 49 | 10 | 42 |
| 8 | 0 | 19 | 2 | 17 |
| 9 | 1 | 14 | 0 | 15 |
| 10 | 2 | 26 | 2 | 27 |
| 11 | 0 | 19 | 2 | 18 |
| 12 | 0 | 12 | 1 | 11 |
| 13 | 0 | 24 | 5 | 19 |
| 14 | 2 | 10 | 2 | 11 |
| 15 | 0 | 14 | 11 | 3 |
| 16 | 0 | 53 | 4 | 48 |
| 17 | 0 | 20 | 0 | 20 |
| 18 | 0 | 21 | 0 | 21 |
| 19 | 1 | 50 | 1 | 48 |
| 20 | 0 | 13 | 1 | 13 |
| 21 | 0 | 13 | 1 | 13 |
| 22 | 0 | 21 | 0 | 21 |

## 7.3   Exact Significance Levels

The preceding result is not an isolated one. Asymptotic approximations are to be avoided except with very large samples. Table 7.9 contains data on oral lesions in three regions of India derived from Gupta et al. [1980] by Mehta and Patel. We want to test the hypothesis that the location of oral lesions is unrelated to geographical region. Possible test statistics include Freeman-Halton $p$ (see Section 7.4), $p_\chi$ and $p_L$. This latter statistic is based on the log-likelihood ratio $\sum\sum f_{ij} \log[f_{ij} f_{..}/f_{i.} f_{.j}]$.

We may calculate the exact significance levels of these test statistics by deriving their permutation distributions or use asymptotic approximations obtained from tables of the chi-square statistic. Table 7.10 taken from the StatXact manual compares the various approaches.

The exact significance level varies from 1% to 3.5% depending on which test statistic we select. Tabulated $p$-values based on large-sample approximations vary from 11% to 23%. Using the Freeman-Halton statistic, the permutation test tells

Table 7.9. Oral Lesions in Three Regions of India

| Site of Lesion | Kerala | Gujarat | Andh |
|----------------|--------|---------|------|
| Labial Mucosa  | 0      | 1       | 0    |
| Buccal Mucosa  | 8      | 1       | 8    |
| Commissure     | 0      | 1       | 0    |
| Gingiva        | 0      | 1       | 0    |
| Hard Palate    | 0      | 1       | 0    |
| Soft Palate    | 0      | 1       | 0    |
| Tongue         | 0      | 1       | 0    |
| Floor of Mouth | 1      | 0       | 1    |
| Alveolar Ridge | 1      | 0       | 1    |

Table 7.10. Three Tests of Independence

| Statistic | $\chi^2$ | F-H | LR |
|-----------|----------|-----|-----|
| Exact $p$-value | 0.0269 | 0.0101 | 0.0356 |
| Tabulated $p$-value | 0.1400 | 0.2331 | 0.1060 |

us the differences among regions are significant at the 1% level; the large-sample approximation says no, they are insignificant even at the 20% level. The permutation test is correct. The large-sample approximation is grossly in error. With so many near-zero entries in the original contingency table, the chi-square large-sample approximation is not appropriate.

## 7.4 Unordered $r \times c$ Contingency Tables

With a computer at hand, the principal issue in the analysis of a contingency table with more than two rows and two columns is deciding on an appropriate test statistic. Halter [1969] showed we can find the probabilities of any individual $r \times c$ contingency table through a straightforward generalization of the hypergeometric distribution given in Equation 7.1. An $r \times c$ contingency table consists of a set of frequencies

$$\{f_{ij}, \ 1 \le i \le r; \ 1 \le j \le c\}$$

with row marginals $\{f_{i.}, \ 1 \le i \le r\}$ and column marginals $\{f_{.j}, \ 1 \le j \le c\}$. Suppose once again we have mixed up the labels. To make matters worse, this time every item/subject is to be assigned both a row label ($f_{1.}$ of which are labeled row 1, $f_{2.}$ of which are labeled row 2 and so forth) and a column label. Let $P$ denote the probability with which a specific table assembled at random will have these exact frequencies.

$$P = Q/R \quad \text{with}^4$$

---

[4] $\prod_{i=1}^{n} f_i! = f_1! f_2! \cdots f_n!.$

$$Q = \prod_{i=1}^{r} f_{i.}! \prod_{j=1}^{c} f_{.j}! f_{..}! \quad \text{and} \quad R = \prod_{i=1}^{r} \prod_{j=1}^{c} f_{ij}!$$

An obvious extension of Fisher's exact test is the Freeman and Halton [1951] test based on the proportion $p$ of tables for which $P$ is greater than or equal to $P_0$ for the original table.

---

### Do It Yourself?

You could write a computer program to perform the tests described in this chapter, one that would select from the set of all tables with a given set of marginals to generate the permutation distribution, but there's a more efficient way. Branch-and-bound algorithms developed by Mehta and Patel [1980, 1983] use a network approach to enumerate only those tables that have a more extreme value of the test statistic than the original. An outline of their method is given in Section 13.5 of Good [2000]. StatXact is the only version of these algorithms that is commercially available at present, and in this chapter, you learn how to use StatXact to do each of the needed tasks.

---

While the extension itself may be obvious, it's not as obvious this extension offers any protection against the alternatives of interest. Just because one table is less likely than another under the null hypothesis does not mean it is going to be more likely under the alternatives of interest to us. Consider the $1 \times 3$ contingency table | $f_1$ | $f_2$ | $f_3$ |, which corresponds to the multinomial with probabilities $p_1 + p_2 + p_3 = 1$; the table whose entries are 1, 2, 3 argues more in favor of the null hypothesis $p_1 = p_2 = p_3$ than of the ordered alternative $p_1 > p_2 > p_3$.

The classic statistic for independence in a contingency table with $r$ rows and $c$ columns is the Pearson chi-square statistic

$$\chi^2 = \sum_{i=1}^{r} \sum_{j=1}^{c} (f_{ij} - Ef_{ij})^2 / Ef_{ij},$$

where $Ef_{ij}$ is the number of observations in the $ij$th category one would expect on theoretical grounds.

With very large samples, this statistic has the chi-square distribution with $(r - 1)(c - 1)$ degrees of freedom. But in most practical applications, the chi-square distribution is only an approximation to the distribution of this statistic and is notoriously inexact for small and unevenly distributed samples.

The permutation statistic based on the proportion $p_\chi$ of tables for which $\chi^2$ is greater than or equal to $\chi_0^2$ for the original table provides an exact test and possesses all the advantages of the original chi-square. The distinction between the two approaches, as we observed in Chapter 6, is that with the original chi-square we look up the significance level in a table, while with the permutation statistic, we derive the significance level from the permutation distribution. With large samples, the two approaches are equivalent, as the permutation distribution converges to the tabulated distribution.[5]

This permutation test has one of the original chi-square test's disadvantages: while it offers global protection against a wide variety of alternatives, it offers no particular protection against any single one of them. The statistics $p$ and $p_\chi$ treat row and column categories symmetrically and no attempt is made to distinguish between cause and effect. To address this deficiency, Goodman and Kruskal [1954] introduce an asymmetric measure of association for nominal scale variables called tau $\tau$ which measures the proportional reduction in error obtained when one variable, the "cause" or independent variable, is used to predict the other, the "effect" or dependent variable.

Assuming the independent variable determines the row,

$$\tau = \frac{\sum_j f_{mj} - f_{m.}}{f_{..} - f_{m.}}$$

where $f_{mj} = \max_i f_{ij}$ and $f_{m.} = \max_i f_{i.}$. $0 \leq \tau \leq 1$. $\tau = 0$ when the variables are independent; $\tau = 1$ when for each category of the independent variables all observations fall into exactly one category of the dependent. These points are illustrated in the following $2 \times 3$ tables:

| 3 | 6 | 9 |
|---|----|----|
| 6 | 12 | 18 |

$\tau = 0$

| 18 | 0 | 0 |
|----|----|---|
| 0 | 36 | 0 |

$\tau = 1$

| 3 | 6 | 9 |
|----|----|---|
| 12 | 18 | 6 |

$\tau = 0.166$

A permutation test of independence is based upon the proportion of tables $p_\tau$ for which $\tau \geq \tau_0$.

Cochran's Q provides an alternate test for independence. Suppose we have $I$ experimental subjects on each of whom we administer $J$ tests. Let $y_{ij} = 1$ or $0$ denote the outcome of the $j$th test on the $i$th patient, e.g., if the test is positive, $y_{ij} = 1$ and is 0 otherwise. Define $R_i = \sum_j y_{ij}$; $C_j = \sum_i y_{ij}$;

$$Q = \frac{\Sigma(C_j - C_.)^2}{R_. - \Sigma(R_i)^2}.$$

## 7.4.1   Causation Versus Association

A significant value for any of the above statistics only means that the variables are associated. It does not mean that there is a cause and effect relationship between them. They may both depend on a third variable omitted from the study.

---

[5] See Chapter 14 of Bishop, Fienberg, and Holland [1975].

---

**Which Test?**

The data are in categories.
The categories can't be ordered.
There are exactly two rows and two columns.
    Use Fisher's Exact Test.

There are more than two rows and at least two columns.
You want to test whether the relative frequencies are the same in each row and in each column.
    Use the Freedman-Halton Test or use chi-square.

You want to test whether the column frequencies depend on the row.
    Use Tau or Q.

---

   Regretably, the converse is also true. A third omitted variable may also result in two variables appearing to be dependent when the opposite is true. Consider the following table, an example of what is termed Simpson's paradox:

|        | Population |         |
|--------|------------|---------|
|        | control    | treated |
| Alive  | 6          | 20      |
| Dead   | 6          | 20      |

We don't need a computer program to tell us the the treatment has no effect on the death rate. Or does it? Consider the following two tables that result when we examine the males and females separately:

|        | Males   |         |
|--------|---------|---------|
|        | control | treated |
| Alive  | 4       | 8       |
| Dead   | 3       | 5       |

|        | Females |         |
|--------|---------|---------|
|        | control | treated |
| Alive  | 2       | 12      |
| Dead   | 3       | 15      |

In the first of these tables, treatment reduces the male death rate from 0.43 to 0.38. In the second from 0.6 to 0.55. Both sexes show a reduction, yet the combined population does not. Resolution of this paradox is accomplished by avoiding a kneejerk response to statistical significance when association is involved. One needs to think deeply about underlying cause and effect relationships before analyzing data. Thinking about cause and effect relationships in the preceding example might have led us to thinking about possible sexual differences, and to employing the analytical techniques described in Section 7.2.1.[6]

---

[6]See also Chapter 8 for a discussion of how to utilize cause and effect relationships in the design

## 7.5    Ordered Statistical Tables

When data are measured on a continuous basis, such as 1.12, 1.13, 1.14, ties are a relatively infrequent occurrence. But when we ask someone to provide a self-rating on a discrete ordinal scale, 1 through 5, for example, ties are inevitable, the rule, not the exception and the methods of this chapter may be more appropriate for analyzing such ordinal data than those of Chapter 3.

Table 7.11. Data Gathered by Graubard and Korn [1987].

|            | Maternal Alcohol Consumption (drinks/day) | | | | | |
|------------|-------|-------|------|------|------|-------|
| Malformation | 0   | < 1   | 1–2  | 3–5  | ≥ 6  | Total |
| Absent     | 17066 | 14464 | 788  | 126  | 37   | 32481 |
| Present    | 48    | 38    | 5    | 1    | 1    | 93    |
|            | 17114 | 14502 | 793  | 127  | 38   | 32574 |

### 7.5.1    Ordered 2 × c Tables

The analysis of a $2 \times c$ ordered contingency table is straightforward and parallels the approach used in Section 3.4 for a $k$-sample comparison, once we have determined what value to assign each of the ordered categories. We illustrate this dilemma with data gathered by Graubard and Korn [1987].

Recall our test statistic is $\sum g[j] f_{1j}$ where $g$ is any monotone increasing function. Among the leading choices for a scoring method $g$ are

i) the category number: 1 for the 1st category, 2 for the second and so forth,

ii) the midrank scores,

iii) scores determined by the user, the choice we made in Section 3.4 when we analyzed the micronuclei data.

Consider the following $1 \times 2$ contingency table

|            | Alcohol Consumption | |
|------------|------|------|
| drinks/day | 0    | 1–2  |
| frequency  | 3    | 5    |

The category or equidistant scores are 1 and 2. The ranks of the 8 observations are 1 through 3, and 4 through 8, so that the midrank score of those in the first category is 2, and in the second 6. Our user-chosen scores, corresponding to alcohol consumption, are 0 and 1.5.

---

of experiments and Section 10.6.1. for further examples of the need for caution when interpreting association as causation.

Table 7.12. Response to Chemotherapy

|  | None | Partial | Complete |
|---|---|---|---|
| CTX | 2 | 0 | 0 |
| CCNU | 1 | 1 | 0 |
| MTX | 3 | 0 | 0 |
| CTX+CCNU | 2 | 2 | 0 |
| CTX+CCNU+MTX | 1 | 1 | 4 |

Using Testimate™ to analyze the full Gaubard-Korn data set, we obtain $p$-values that range from the nonstatistically significant—0.29 for midrank scores, to marginally significant—0.10 for the equidistant scores, to highly significant, 0.01 for our user-chosen scores. A user-chosen score based on the user's knowledge of underlying cause and effect is always recommended as it will be the most effective at distinguishing between hypothesis and alternative.

## 7.5.2    More Than Two Rows and Two Columns

Two cases need be considered. The first when the columns but not the rows of the table may be ordered (the other variable being purely categorical), and the second when both columns and rows can be ordered.

### Singly Ordered Tables

Our approach parallels that of Section 8.2 in which we describe a $k$-sample comparison of metric data. Our test statistic is

$$F_2 = \sum (T_i - \overline{T})^2 \qquad \text{where} \qquad T_i = \sum g_j f_{ij}.$$

As in the case of the $2 \times c$ table our problem is in deciding on the appropriate scores $\{g_j\}$.

Table 7.12 provides tumor regression data for five chemotherapy regimes. As partial response corresponds to approximately two years in remission (about 100 weeks) and complete response to an average of three years (150 weeks), we assign scores of 0, 100, and 150 to the ordered response categories.

To use StatXact to analyze the tumor data, we first click on TableData in the main menu, click on Settings, then enter Column Scores. To execute the analysis, we select in turn Statistics, Singly Ordered $R \times C$ Table, ANOVA with Arbitrary Scores, and Exact test method. The results are tabulated below.

Although our estimated significance level is less than 0.0444, a 99% confidence interval for this estimate, based on a sample of 3200 possible rearrangements, does include values greater than 0.05. We can narrow this confidence interval by sampling additional rearrangements. With a Monte Carlo of 10,000 sample tables, our estimate of the p-value is 0.0434 with a 99% confidence interval of (0.0382, 0.0486). Of course, the calculations take three times as long. When I

---

**ANOVA TEST [that the 5 rows are identically distributed]**

```
Datafile: C:\EXAMPLES\TUMOR.CY3
Statistic based on the observed data :
  The Observed Statistic = 7.507
Asymptotic p-value: (based on chi-square distribution with 4 df )
  Pr { Statistic .GE. 7.507 } = 0.0747
Monte Carlo estimate of p-value :
  Pr { Statistic .GE. 7.507 } = 0.0444
  99.00% Confidence Interval = ( 0.0350, 0.0538 )
```

---

first perform a permutation test, I use as few as 400 to 1600 simulations. If the results are equivocal as they were in this example, then and only then will I run 10,000 simulations.

### Doubly Ordered Tables

In an $r \times c$ contingency table conditioned on fixed marginal totals, the outcome depends on the $(r-1)(c-1)$ odds ratios

$$\phi_{ij} = \frac{\pi_{ij}\pi_{i+1,j+1}}{\pi_{i,j+1}\pi_{i+1,j}},$$

where $\pi_{ij}$ is the probability of an individual being classified in row $i$ and column $j$.

In a $2 \times 2$ table, conditional probabilities depend on a single odds ratio and hence one- and two-tailed tests of association are easily defined. In an $r \times c$ table there are potentially $n = 2(r-1)(c-1)$ values, two for each of the $(r-1)(c-1)$ odds ratios. Hence, an omnibus test for no association, e.g., $\chi^2$, might have as many as $2^n$ tails.

Following Patefield [1982], we consider tests of the null hypothesis of no association between row and column categories $H : \phi_{ij} = 1$ for all $i$, $j$ against the alternative of a positive trend $K : \phi_{ij} \geq 1$ for all $i$, $j$.

The principal test statistic considered by Patefield [1982], also known as the linear-by-linear association test, is

$$\lambda = \sum \sum f_{ij} r_i c_j$$

where $\{r_i\}$ and $\{c_j\}$ are user-chosen row and column scores.[7]

## 7.6  Summary

In this chapter, you were introduced to the concept of a contingency table with fixed marginals, and shown you could test against a wide variety of general and

---

[7]This statistic is actually just another form of Mantel's $U$, perhaps the most widely used of all multivariate statistics. For a worked-through example, see Section 9.3.2.

specific alternatives by examining the resampling distribution of the appropriate test statistic. Among the test statistics you considered were Fisher's Exact, Freedman–Halton, Pearson's Chi-Square, Tau, Q, Pitman's correlation, and
linear-by-linear association. These latter two statistics are to be used when you
could take advantage of an ordering among the categories.

## 7.7  To Learn More

Excellent introductions to the analysis of contingency tables may be found in
Agresti [1990, 1992], and in the StatXact manual authored by Mehta and Patel
(see Appendix 3). Major advances in analysis by resampling means have come
about through the efforts of Gail and Mantel [1977], Mehta and Patel [1983],
Mehta, Patel, and Senchaudhuri [1988], Baglivo, Oliver, and Pagano [1988], and
Smith, Forester, and McDonald [1996].

Berkson [1978], Basu [1979], Haber [1987] and Mielke and Berry [1992] examine Fisher's exact test. The power of the Freeman–Halton statistic in the $r \times 2$
case is studied by Krewski, Brennan, and Bickis [1984]. Details of the calculation of the distribution of Cochran's Q under the assumption of independence are
given in Patil [1975]. For a description of some other, alternative statistics for use
in $r \times c$ contingency tables, see Nguyen [1985].

To study a $2 \times 2$ table in the presence of a third covariate, see Bross [1964] and
Mehta, Patel, and Gray [1985].

## 7.8  Exercises

1. Do the four hospitals considered in Exercise 16 of Chapter 1 have the same
   billing practices?

2. A total of 40,500 babies born in 1981 in the United States died before they
   were 28 days old. Of these babies, 30,000 were white, and 10,500 were
   nonwhite. Comment on the hypothesis that black kids have a better chance
   to survive in North America than white kids.

3. Your friend rolls a die 120 times. Each time, before she rolls, you concentrate
   and visualize the die coming up a six. The die actually lands six a total of 28
   times. Are you psychic?

4. A preliminary poll conducted well before the 1996 U.S. presidential election
   yielded the following results:

   |         | Males | Females |
   |---------|-------|---------|
   | Dole    | 65    | 35      |
   | Clinton | 55    | 42      |

Are these differences significant?

5. Referring to Table 7.8, if Sandoz excluded cite 15 from their calculations, could they safely combine the data from the remaining cites?

6. Will encouraging your child promote his or her intellectual development? A sample of 100 children and their mothers were observed and the children's IQs tested at 6 and 12 years. Results were as follows:

| | Mothers Encourage Schoolwork | | |
|---|---|---|---|
| | rarely | sometimes | always |
| IQ increased | 8 | 15 | 27 |
| IQ decreased | 30 | 9 | 11 |

   a. Do you plan to perform a one-tailed or two-tailed test?
   b. What is the significance level of your test?

7. Does 1,2 dichloro-ethane induce tumors? Consider the following data evaluated by Gart et al. [1986].

| | Tumor | No Tumor |
|---|---|---|
| Treated | 15 | 21 |
| Control | 2 | 35 |

8. Recall from Section 4.5 that while my college team McGill won 11 of its 15 games last season, your team only won 8 of its 14. Can you use Fisher's Exact test to analyze this data? Would your answer be the same under all circumstances? What if McGill and your team compete in the same league?

9. Referring to the literature of your own discipline, see if you can find a case where an $r \times 2$ table with at least one entry smaller than 7 gave rise to a border line $p$-value using the traditional chi-square approximation. Reanalyze this table using resampling methods. Did the authors use a one-tailed or a two-tailed test? Was their choice appropriate?

10. Show that once you have selected $(r-1)(c-1)$ of the entries in a contingency table with $r$ rows and $c$ columns the remainder of the entries are determined.

11. Holmes and Williams [1954] studied tonsil size in children to verify a possible association with the virus *S. pyrogenes*. Do you feel there is an association? How many rows and columns in the following contingency table? Which, if any, of the variables is ordered?

| | Tonsil Size by Whether Carrier of *S. pyrogenes* | | |
|---|---|---|---|
| | Not Enlarged | Enlarged | Greatly Enlarged |
| Noncarrier | 497 | 560 | 269 |
| Carrier | 19 | 29 | 24 |

**12.** Is treatment A superior? Hints: Are the categories ordered? Is this a one- or two-tailed test?

| Outcome | A | B |
|---|---|---|
| Improvement | | |
| marked | 7 | 3 |
| moderate | 15 | 9 |
| slight | 16 | 14 |
| No change | 13 | 21 |
| Worse | 1 | 5 |

**13.** Does dress make the woman?

| | Time to First Promotion (months) | | |
|---|---|---|---|
| | Long | Average | Short |
| Poorly dressed | 12 | 8 | 4 |
| Well dressed | 18 | 25 | 20 |
| Very well dressed | 13 | 19 | 31 |

**14.** Do Supreme Court appointments follow a Poisson distribution? Actual United States Supreme Court appointments grouped in five-year periods from 1800 to 1990 appear below:

| Period | Actual Poisson | Period | Actual Poisson |
|---|---|---|---|
| 1800-04 | 2 | 1900-04 | 2 |
| 1805-09 | 2 | 1905-09 | 2 |
| 1810-14 | 2 | 1910-14 | 6 |
| 1815-19 | 0 | 1915-19 | 2 |
| 1820-24 | 1 | 1920-24 | 4 |
| 1825-29 | 2 | 1925-29 | 1 |
| 1830-34 | 1 | 1930-34 | 3 |
| 1835-39 | 5 | 1935-39 | 4 |
| 1840-44 | 1 | 1940-44 | 5 |
| 1845-49 | 3 | 1945-49 | 4 |
| 1850-54 | 2 | 1950-54 | 1 |
| 1855-59 | 1 | 1955-59 | 4 |
| 1860-64 | 5 | 1960-64 | 2 |
| 1865-69 | 0 | 1965-69 | 3 |
| 1870-74 | 4 | 1970-74 | 3 |
| 1875-79 | 1 | 1975-79 | 1 |
| 1880-84 | 4 | 1980-84 | 1 |
| 1885-89 | 3 | 1985-89 | 2 |
| 1890-94 | 4 | 1990-94 | 4 |
| 1895-99 | 2 | 1995-99 | |

Recall from Section 4.2, that if an observation $X$ has a Poisson distribution, such that we may expect an average of $\lambda$ events per interval, the probability that $X = k$ in a given interval is $k^{-1} \exp[-\lambda k]$ for $k = 0, 1, 2, \ldots$. To test the hypothesis that Supreme Court appointments have the Poisson distribution, proceed as follows:

1) Compute the arithmetic average and use it to estimate $\lambda$.

2) Use the Poisson formula to fill in the remaining column of the table.

3) Use a statistics program to determine the probability of observing a $2 \times k$ contingency table that is as or more extreme.

4) Would you use the same test to establish whether Supreme Court appointments are made once every two years on the average?

# CHAPTER 8

# Experimental Design and Analysis

Failing to account for or balance extraneous factors can lead to major errors in interpretation. In this chapter, you learn to block or measure all factors that are under your control and to utilize random assignment to balance the effects of those you cannot. You learn to design experiments to investigate multiple factors simultaneously, thus obtaining the maximum amount of information while using the minimum number of samples.

## 8.1 Noise in the Data

In Section 2.6, we encountered a woman anxious to demonstrate her abilities as a tea taster. We argued in that chapter that by removing all extraneous sources of variation—e.g., the appearance of the cup, the temperature of the water, the expression on the experimenter's face, we could focus more narrowly on the factor to be tested.

Eliminating or reducing extraneous variation is the first of several preventive measures we use each time we design an experiment or survey. We strive to conduct our experiments in a biosphere with atmosphere and environment totally under our control. And when we can't—which is almost always the case—we record the values of the extraneous variables to use them either as *blocking* units or as *covariates*.

### 8.1.1 Blocking

Although the significance level of a permutation test may be distribution free, its power (defined in Section 6.1.3) strongly depends on the underlying distribution.

In Figure 8.1, we see that the more variable our observations, the less the power of our tests and our ability to detect the alternative. One way to reduce the variance is to *block* the experiment, that is, to subdivide the population into more homogeneous subpopulations and to take separate independent samples from each.

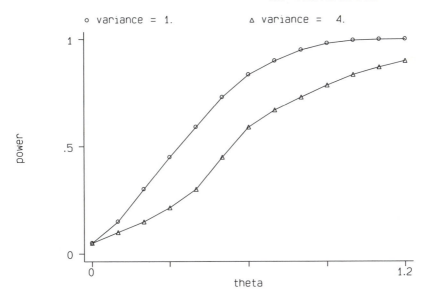

**Figure 8.1.**   Power as a function of the population variance.

Suppose you were designing a survey on the effect of income level on the respondent's attitude toward compulsory pregnancy. Obviously, the views of men
and women differ markedly on this controversial topic. To reduce the variance and
increase the power of your tests, *block* the experiment, interviewing and reporting
on men and women separately. A physician would want to block by gender in a
medical study, and probably by age and race as well. An agronomist would want
to distinguish among clay soil, sand, and sandy-loam.

Whenever a population can be subdivided into distinguishable subpopulations,
you can reduce the variance of your observations and increase the power of your
statistical tests by blocking or stratifying your sample.

Suppose we have agreed to divide our sample into two blocks—one for men,
one for women. If this is an experiment, rather than a survey, we would then
assign subjects to treatments separately and independently within each block. In
a study that involves two treatments and 10 experimental subjects, four men and
six women, we would first assign the men to treatment and then the women. We
could assign the men in any of $\binom{4}{2} = 6$ ways and the women in any of $\binom{6}{3} = 20$
ways, for a total of $6 \times 20 = 120$ possible random assignments in all.

When we come to analyze the results of our experiment, we use the permutation
approach to ensure that we analyze in the way the experiment was designed. Our
test statistic is a natural extension of that used for the two-sample comparison;
we simply sum up the separate sums computed for each block

$$S = \sum_{b=1}^{B} \sum_{j=1}^{n_b} x_{bj},$$   (8.1)

---

**How Many Controls?**

Every experiment involves at least two groups of subjects—those that took the drug and those who took the placebo (the control group), or those that took the new drug versus those that took the old.

For the self-assured, the smug, and the person with something to sell, controls may be an unnecessary luxury. But the day I started at the Upjohn Company, my boss John R. Schultz recommended I use twice as many subjects in each control group as the number devoted to experimental treatment, and that I use two types of controls, positive and negative.

For a study on Motrin, the positive control was aspirin (the best treatment at the time) and the negative or neutral control was a harmless filler used to give aspirin tablets their compact shape. The reason we used so many subjects in each control group? "Life is full of surprises," said John, "you leave work one day whistling, the next day you're back with a head cold. Most of the time these negative effects have nothing to do with the treatment. By using many control subjects, you ensure the normal wear and tear of ordinary life will be detected and accounted for and *won't* be falsely associated with the treatment you're trying to investigate."

Further proof of John's wisdom came several years later with the controversy over silicon implants. Women who had the implants suffered from a wide variety of complaints. Dow Corning, the manufacturer, didn't have the data on control groups to show that such complaints might be pure coincidence and was forced to pay millions of dollars in compensation.

---

where $B$ is the number of blocks, two in the present example, and the inner sum extends over the $n_b$ treated observations $\{x_{bj}\}$, within each block.[1]

We compute the test statistic for the original data. Then, we rearrange the observations at random independently within each block, subject to the restriction that the number of observations within each treatment category—in this example, the pair $\{n_b, m_b\}$—remains constant.

We compute $S$ for each of the 120 possible rearrangements. If the value of $S$ for the original data is among the $120\alpha$ largest values, then we reject the null hypothesis at the $\alpha$ significance level; otherwise we accept it.

The resulting permutation test is exact and most powerful against normal alternatives even if the observations on men and women have different distributions (Lehmann, 1986). As we saw in Section 6.2, all that is required is the subsets of errors be exchangeable.

The design need not be balanced. The test statistic $S$ (Equation 8.1) is a sum of several independent sums. Unequal sample sizes resulting from missing data or an inability to complete one or more portions of the experiment will affect the analysis only in the relative weights assigned to each subgrouping.[2]

---

[1] Recall that $\sum_{k=1}^{n} X_k = X_1 + X_2 + \cdots + X_n$ where $X_i$ is the $i$th observation.

[2] Warning: This remark applies only if the data is missing at random. If treatment-related with-

---

**Analyzing a Blocked Experiment**

Compute the test statistic for the original data, for example,

$$S_0 = \sum_{b=1}^{B} \sum_{j} x_{bj},$$

where the inner sum extends over the first sample in the block.
Repeat 400 times:
   For each block:
      Rearrange the data in the block.
      Compute $S$.
      Record the number of times $S$ *exceeds* $S_0$.
   If the proportion of rearrangements in which $S$ exceeds $S_0$ is small, conclude
that the difference between the samples is statistically significant.

---

Blocking is applicable to any number of subgroups; the extreme case in which
every pair of observations forms a distinct subgroup is *matched pairs*, which we
studied in 3.7.

## 8.1.2  Measuring Factors We Can't Control

Many of the factors in an experiment will be beyond our control. Rainfall in an
agricultural experiment is one example. If we are studying the effects of advertising on the sale of beer, a sudden heat wave or an unexpected drop in temperature
will markedly affect our results. But we can measure rainfall and temperature,
and use our observations either to block or to correct our results.

We could categorize rainfall as light, normal, or heavy and, as in Section 2.2,
incorporate rainfall in some kind of model. For example, if $X_i$ denotes the yield
of one of our test acres, we might want to consider the corrected yield $X'_i = X_i - a - bR_i$, where $R_i$ is the quantity of rain that fell on that acre, and $a$ and $b$
are constants in our model. We would then apply our permutation procedures to
the corrected values.

## 8.1.3  Randomization

Imagine you are all prepared for an experiment to find a new and better fertilizer for growing tomatoes. You've set up the experiment indoors under the lights
so you can keep total control over duration of daylight, moisture, and temperature. Since you are hunting for a general-purpose fertilizer, you've put out three
separate sets of trays containing sandy loam, a sand-clay mix, and a clay soil respectively. Now, it just remains to put the plants in the soil. Careful . . . Tomato

---

drawals are a problem in one of your studies, see Entsuah [1990] for the details of a resampling
procedure.

---

**Experimental Design**

LIST factors you feel may influence the outcome of your experiment.
BLOCK factors under your control. You may want to use some of these factors
to restrict the scope of your experiment, e.g., eliminate all individuals under 18
and over 60.
MEASURE remaining factor.
RANDOMLY assign units to treatment within each block.

---

plants are not all alike and there are big differences among seedlings. Studies
have shown there is a natural tendency to plant the tall seedlings first and save the
runts for last. This is not a good idea if all the runts end up in one tray and all the
big sturdy plants in another.

An alternate plan would be to deal the seedlings out like cards, the first for
tray 1, the second for tray 2, and so forth down the line. Alas, this method would
still seem to put the better plants into the first tray.

When we analyze this experiment using permutation methods, we assume that
each assignment of labels to treatments is equally likely. Only if we assign the
plants to the different trays completely at random, will this assumption be ful-
filled. If the plants assigned to tray 1 prove in the end to be larger on the average
than those in tray 2, this should be strictly a matter of chance, not the result of a
specific bias on our part.

*Randomly assign subjects to treatment whenever you don't or can't control all
the factors in an experiment.*

Our tomato study might use one of the following randomization schemes:

1. Choose a random rearrangement of the integers, $1 \ldots m$, corresponding to
   the $m$ trays. Place the first $m$ plants in the trays in the order specified.
   Choose a second random rearrangement, a third, and so on, until all the
   plants have been placed in the trays.

2. Throw an $m$-sided die and place the first plant in the indicated tray. Continue to throw the die until all the plants have been placed. If the die indicates that an already full tray has been selected, just ignore it and throw
   again. (See also Exercise 4.)

## 8.2   $k$-Sample Comparison

Suppose we wanted to assess the effect on crop yield of hours of sunlight, observing the yield $X_{ij}$ for $I$ different levels of sunlight, $i = 1, \ldots, I$ with $n_i$
observations at each level. Our model is that

$$X_{ij} = \mu_i + \epsilon_{ij},$$

where the experimental errors $\{\epsilon_{ij}\}$ all come from the same distribution.

Let $X_{i.} = \sum_{j=1}^{n_i} X_{ij}/n_i$ and $X_{..} = \sum_{i=1}^{I} \sum_{j=1}^{n_i} X_{ij}/N$. The sum of squares of the deviations about the grand mean may be analyzed into two sums,

$$\sum_{i=1}^{I} \sum_{j=1}^{n_i} (X_{ij} - X_{..})^2 = \sum_{i=1}^{I} \sum_{j=1}^{n_i} (X_{ij} - X_{i.})^2 + \sum_{i=1}^{I} n_i (X_{i.} - X_{..})^2,$$

the first of which represents the within-level sum of squares and the second the between-level sum of squares.[3]

For testing the hypothesis that the level means are all the same, against the alternative that at least one mean is different from the others, only the between-level sum is of interest. The grand mean $X_{..}$ is the same for all permutations. Only the sum $F_2 = \sum_i n_i (X_{i.})^2 = \sum_i (\sum_j X_{ij})^2/n_i$ changes as we rearrange the observations among levels and it is this sum whose permutation distribution we use to test the hypothesis.

Alternatively, we might use the statistic $F_1 = \sum_i n_i |X_{i.} - X_{..}|$. Whether we use $F_1$ or $F_2$ will depend on our loss function (see Section 6.1.4) and the relative weight we assign to far versus near alternatives. $F_2$ would be our choice, for example, when observations are expensive or difficult to obtain and we want to increase the probability of detecting distant alternatives even if it means decreasing the probability of detecting near ones. We may also want to substitute medians for means, or replace the original observations by their ranks or some other transformation [see Appendix 2].

The statistics $F_2$ and $F_1$ offer protection against a broad variety of shift alternatives including

$$K_1 : s_1 = s_2 > s_3 = \cdots$$
$$K_2 : s_1 > s_2 > s_3 = \cdots$$
$$K_3 : s_1 < s_2 > s_3 = \cdots$$

As a result, they may not provide a most powerful test against any alternative. If we believe the effect to be ordered and monotone increasing as in "more sunlight means more growth," then we should use the Pitman correlation introduced in Section 3.4, $P = \sum_i g[i] \sum_j X_{ij}$, to test the hypothesis.

---

[3] The dot . used as a subscript indicates we have summed over the corresponding subscript; thus

$$X_{ijk.} = \sum_{m=1}^{M} X_{ijkm}.$$

The bar over the letter indicates we have taken the average; thus

$$\overline{X}_{ijk.} = \frac{1}{M} \sum_{m=1}^{M} X_{ijkm}.$$

Table 8.1. Effects of Fertilizer on Crop Yield

| FastGro | NewGro | WunderGro |
|---------|--------|-----------|
| 27      | 38     | 75        |
| 30      | 12     | 76        |
| 55      | 72     | 54        |
| 71      |        |           |
| 18      |        |           |

Table 8.2. Effects of Fertilizer on Crop Yield; Deviations from Grand Mean

| FastGro | NewGro | WunderGro |
|---------|--------|-----------|
| $-21$   | $-10$  | $+27$     |
| $-18$   | $-36$  | $+28$     |
| $+07$   | $+24$  | $+06$     |
| $+23$   |        |           |
| $-30$   |        |           |

## 8.2.1   Analyzing a One-Way Table

Suppose we wish to compare the effects of three brands of fertilizer on crop yield as recorded in Table 8.1.

We'll use $F_1$ as our test statistic which in this example *is*

$$5|\overline{X}_{1.} - \overline{X}_{..}| + 3|\overline{X}_{2.} - \overline{X}_{..}| + 3|\overline{X}_{3.} - \overline{X}_{..}|.$$

$\overline{X}_{..} = 48$; to simplify our calculations, we'll first subtract 48 from each observation as in Table 8.2. $F_1$ reduces to the sum of the absolute values of the sums of these deviations or $39 + 22 + 61 = 122$.

In Table 8.3, we have rearranged the observations, preserving the number of observations in each category and the new value of $F_1$ is $2 + 23 + 25 = 50$.

Continuing in this fashion until we have considered all possible rearrangements of the observations among the categories, $\binom{11}{5\,3}$, we see a value as large as 122 occurs less than 1% of the time, and conclude there is a statistically significant difference among the fertilizers.[4]  A C++ program to do these calculations is included in Appendix 2.

## 8.3   Balanced Designs

Now suppose we want to assess the simultaneous effects on crop yield of hours of sunlight and rainfall. We determine to observe the crop yield $X_{ijm}$ for $I$ different levels of sunlight, $i = 1, \ldots, I$, and $J$ different levels of rainfall, $j = 1, \ldots, J$,

---

[4]Recall from Section 2.5.1 that $\binom{11}{5\,3}$ means $\frac{11!}{5!3!(11-8)!}$.

Table 8.3. Effects of Fertilizer on Crop Yield; Rearranged Deviations

| FastGro | NewGro | WunderGro |
|---------|--------|-----------|
| −21 | −10 | +27 |
| −18 | −36 | +28 |
| +07 | +23 | −30 |
| +24 | | |
| +06 | | |

---

### Analyzing a One-Way Table

$H$: mean/median the same for all treatments.
$K$: means/medians are different for at least one treatment.

Assumptions:

1. Observations are exchangeable if the hypothesis is true.

2. Choose a test statistic:
   Treatments cannot be ordered and
      Even small differences in means are important—Use $F_1$.
      Only large differences in means are important—Use $F_2$.
   Treatments can be ordered—Use R.

3. Compute the test statistic for the original data.

4. Repeat 400 times:
   Rearrange the data.
   Compute test statistic.
   Record number of times statistic exceeds original value.

---

and to make $M$ observations at each factor combination, $m = 1, \ldots, M$. We adopt as our model relating the dependent variable crop yield (the effect) to the independent variables of sunlight and rainfall (the causes)

$$X_{ijm} = \mu + s_i + r_j + (sr)_{ij} + \epsilon_{ijm}.$$

In this model, terms with a single subscript like $s_i$, the effect of sunlight, are called *main effects*. Terms with multiple subscripts like $(sr)_{ij}$, the residual and nonadditive effect of sunlight and rainfall, are called *interactions*. The *residuals* $\{\epsilon_{ijm}\}$ represent that portion of crop yield that can not be explained by the independent variables alone. To ensure the residuals are exchangeable so that permutation methods can be applied, the experimental units must be assigned at random to treatment.

Suppose we also want to compare the effects of several different fertilizers for different combinations of sunlight and rainfall. We would observe the crop yield $X_{ijkm}$ for $I$ different levels of sunlight, $J$ different levels of rainfall, $K$ different levels of fertilizer, and make $M$ observations at each factor combination. Our model would then be

$$X_{ijkm} = \mu + s_i + r_j + f_k + (sr)_{ij} + (sf)_{ik} + (rf)_{jk} + (srf)_{ijk} + \epsilon_{ijkm}.$$

In this model we have three main effects, $s_i, r_j, f_k$, three two-way interactions, $(sr)_{ij}, (sf)_{ik}, (rf)_{jk}$, a single three-way interaction, $(srf)_{ijk}$, and the error term $\epsilon_{ijkm}$.

Including the additive constant $\mu$ in the model allows us to define all main effects and interactions so they sum to zero, $\sum s_i = 0$; $\sum r_j = 0$; $\sum f_k = 0$; $\sum \sum (sr)_{ij} = 0$; $\sum \sum (sf)_{ik} = 0$; $\sum \sum (rf)_{jk} = 0$; $\sum \sum \sum (srf)_{ijlk} = 0$. As a result, the null hypothesis of no effect of sunlight on crop yield is equivalent to the statement that each of the main effects $s_i = 0$ for $i = 1, \ldots, I$. Under the alternative, the different terms $s_i$ represent deviations from a zero average, with the interaction term $(sr)_{ij}$ representing the deviation from the sum $s_i + r_j$.

When we have multiple factors, we must also have multiple test statistics. In the preceding example, we require three separate tests and test statistics for the three main effects $s_i, r_j$, and $f_k$, plus four other statistical tests for the three two-way and the one three-way interactions. Will we be able to find statistics that measure a single intended effect without *confounding* it with a second unrelated effect? Will the several test statistics be independent of one another?

The answer is yes to both questions only if the design is *balanced*, that is, if there are equal numbers of observations in each subcategory, and if the test statistics are independent of one another. In an unbalanced design, main effects will be confounded with interactions so the two can not be tested separately, a topic we return to with a bootstrap solution in Section 8.6.

### 8.3.1  Main Effects

In a $k$-way analysis with equal sample sizes $M$ in each category, we assess the main effects using essentially the same statistics we would use for randomized blocks. Take sunlight in the preceding example. If we have only two levels of sunlight then, referring to Equation 8.1, our test statistic for the effect of sunlight is

$$S = \sum_j \sum_k \sum_m X_{1jkm}, \qquad (8.2)$$

the sum of all observations at the first level of sunlight.

If we have more than two levels of sunlight, our test statistic is

$$F_2 = \sum_i \sum_j \sum_k (X_{ijkm})^2 \qquad (8.3)$$

or

$$F_1 = \sum_i \sum_j \sum_k |X_{ijkm} - \overline{X}_{.jk.}|. \qquad (8.4)$$

Table 8.4. Effect of Sunlight and Fertilizer on Crop Yield

| | | Lo | Med | High |
|---|---|---|---|---|
| **S** | | | **Fertilizer** | |
| | | Lo | Med | High |
| **U** | **Lo** | 5 | 15 | 21 |
| **N** | | 10 | 22 | 29 |
| **L** | | 8 | 18 | 25 |
| **I** | | | | |
| **G** | **Hi** | 6 | 25 | 55 |
| **H** | | 9 | 32 | 60 |
| **T** | | 12 | 40 | 48 |

If we believe the effect to be monotone increasing, then, in line with the thinking detailed in Section 3.4, we would use the Pitman correlation statistic

$$R = \sum_i \sum_j \sum_k g[i]\overline{X}_{ijk}. \qquad (8.5)$$

To obtain the permutation distributions of the test statistics $S$, $F_2$, $F_1$, and $R$ for the effect of sunlight, we permute the observations independently in each of the $JK$ blocks determined by a specific combination of rainfall and fertilizer. Exchanging observations within a category corresponding to a specific level of sunlight $i$ leaves the statistics $S$, $F_2$, $F_1$, and $R$ unchanged. We can concentrate on exchanges between categories and the total number of rearrangements is $JK\binom{IM}{M\cdots M}$.

We compute the test statistic ($S$, $F_1$, or $R$) for each rearrangement, rejecting the hypothesis that sunlight has no effect on crop yield only if the value of $S$ (or $F_1$ or $R$) that we obtain using the original arrangement of the observations lies among the $\alpha$ most extreme of these values.[5]

## 8.3.2  Analyzing a Two-Way Table

In this second example, we apply the permutation method to determine the main effects of sunlight and fertilizer on crop yield using the data from the two-factor experiment depicted in Table 8.4. As there are only two levels of sunlight in this experiment, we use $S$ (Equation 8.1) to test for the main effect.

For the original observations, $23 + 55 + 75 = 153$. One possible rearrangement is shown in Table 8.5 in which we have interchanged the two observations marked with an asterisk, the 5 and 6. The new value of $S$ is 154.

---

[5]A possible alternative to the statistics $S$, $F_1$, and $F_2$ would appear to be the standard $F$-ratio statistic

$$F = \frac{MJK \sum_i (\overline{X}_{i..} - \overline{X}_{...})^2}{(I-1)\widehat{\sigma}^2},$$

where $\widehat{\sigma}^2$ is our estimate of the variance of the errors $\epsilon_{ijkm}$. But if we use $F$, we are forced to consider exchanges between as well as within blocks, thus negating the advantages of blocking.

Table 8.5. Effect of Sunlight and Fertilizer Data Rearranged

|        |      | Lo    | Med | High |
|--------|------|-------|-----|------|
| **Lo** |      | 6*    | 15  | 21   |
|        |      | 10#   | 22  | 29   |
|        |      | 8     | 18  | 25   |
|        |      |       |     |      |
| **Hi** |      | 5*    | 25  | 55   |
|        |      | 9#    | 32  | 60   |
|        |      | 12    | 40  | 48   |

Table 8.6. Effect of Sunlight and Fertilizer on Crop Yield; Testing for Nonadditive Interaction

|       |        | **Fertilizer** | | |
|-------|--------|------|------|-------|
| **S** |        | Lo   | Med  | High  |
| **U** | **Lo** | 4.1  | −2.1 | −11.2 |
| **N** |        | 9.1  | 4.1  | −3.2  |
| **L** |        | 7.1  | 0.1  | −7.2  |
| **I** |        |      |      |       |
| **G** | **Hi** | −9.8 | −7.7 | 7.8   |
| **H** |        | −7.8 | −0.7 | 12.8  |
| **T** |        | −3.8 | 7.2  | 0.8   |

As can be seen by a continuing series of straightforward hand calculations, the test statistic $S$ or the main effect of sunlight is as small or smaller than it is for the original observations in only 8 out of the $\binom{6}{3}^3 = 8000$ possible rearrangements. For example, it is smaller when we swap the 9 of the Hi-Lo group for the 10 of the Lo-Lo group (the two observations marked with the pound sign #). We conclude the effect of sunlight is statistically significant.

The computations for the main effect of fertilizer are more complicated—we must examine $\binom{9}{3\,3}^2$ rearrangements, and compute the statistic $F_1$ for each. We use $F_1$ rather than $R$ because of the possibility that too much fertilizer–the "High" level–might actually suppress growth. Only a computer can do this many calculations quickly and correctly, so we adapted our C++ program from Appendix 2 to make them. The estimated significance level is 0.001 and we conclude that this main effect, too, is statistically significant.

In this last example, each category held the same number of experimental subjects. If the numbers of observations were unequal, our main effect would have been confounded with one or more of the interactions (see Section 8.5). In contrast to the simpler designs we studied in Chapter 3, missing data will affect our analysis.

### 8.3.3   Testing for Interactions

In the preceding analysis of main effects, we assumed the effect of sunlight was the same regardless of the levels of the other factors. But this may not always be the case. Of what value is sunlight to a plant if there is not enough fertilizer in the ground to support growth? Or vice versa, of what value is fertilizer if there is insufficient sunlight? Sunlight and fertilizer *interact*,[6] we cannot simply add their effects. Our model should and does include an interaction term $(sf)_{ij}$:

$$X_{ijm} = \mu + s_i + f_j + (sf)_{ij} + \epsilon_{ijm}.$$

Suppose we were to eliminate row and column effects by subtracting the row and column means from the original observations as in Table 8.6.[7]

$$X'_{ijk} = X_{ijk} - \overline{X}_{i..} - \overline{X}_{.j.} + \overline{X}_{...}.$$

The pattern of plus and minus signs in this table of residuals suggests that fertilizer and sunlight affect crop yield in a superadditive or synergistic fashion. Note the minus signs associated with the mismatched combinations of a high level of sunlight and a low level of fertilizer and a low level of sunlight with a high level of fertilizer.

An obvious statistic to test the hypothesis that $(sf)_{ij} = 0$ for all $i$ and $j$ would appear to be

$$I = \sum_i \sum_j \left( \sum_k X'_{ijk} \right)^2. \tag{8.6}$$

Unfortunately, the labels on the deviates $X'_{ijk}$ are not exchangeable so we cannot permute them. The expected value of $X'_{ijk}$ is $\epsilon_{ijk} - \overline{\epsilon}_{i..} - \overline{\epsilon}_{.j.} + \overline{\epsilon}_{...}$. Thus the deviates are correlated and the correlation between two deviates in the same row or column has a different value than the correlation between deviates in different rows or columns.

The only exact test for nonzero interactions requires the use of synchronized permutations (Pesarin, 2001, Chapter 8) and is beyond the scope of this text. We are safer in most instances to conclude the two variables do not act independently and in reporting our results to always refer to the *combined* effect of sunlight and fertilizer.

### 8.3.4   A Worked-Through Example

Consider the following example suggested by Tony DuSoir, the author of SC,[8] which is taken from Hettmansperger [1984]. Survival times are measured for each of three different poisons and four different treatments. The experiment is replicated four different times and the results are displayed in Table 8.7.

---

[6]The reinforcing effects of sunlight and fertilizer in this example are said to be *synergistic*. It is easy to think of examples where two effects are *antagonistic*, and still others where the factors are alternately synergistic and antagonistic depending on their levels.

[7]By adding the grand mean, $\overline{X}_{...}$, we ensure the overall sum will be zero.

[8]See Appendix 3.

## Analyzing a Two-Way Table

Interactions:

$H_1$: Treatment effects are the same for all levels of the other factor.
$K_1$: Treatment effect depends upon the level of the other factor.

Assumptions:
  Labels on the errors can be exchanged.

Procedure:

  1. Test using synchronized permutations OR
  2. Compute cell means $\{\overline{X}_{ij.}\}$, row means $\{\overline{X}_{i..}\}$, column means $\{\overline{X}_{.j.}\}$, and grand mean $\overline{X}_{...}$.
     Calculate and display residuals.

  If the resulting test or display suggests an interaction, then you will need to perform a one-way analysis for one of the factors at *each* level of the second factor. Otherwise, proceed to test for main effects.

  Main Effects:
$H_2$: Treatment has no effect.
$K_2$: Outcome depends upon the treatment.

Assumptions:
  Observations are exchangeable if the hypothesis is true.

Procedure:
  For *each* of the treatments
  Choose a test statistic.
      Treatment levels *cannot* be ordered and
          Even small differences in means are important — Use $F_1$.
          Only large differences in means are important — Use $F_2$.
      Treatment levels *can* be ordered — Use R.
  Compute the statistic for the original observations.
  Obtain permutation distribution of the statistic,
      Consider only rearrangements that result in exchanging
      values between treatment levels.
  Draw a conclusion.

Table 8.7. Survival Times Following Treatment

|   |   | Treatment | | | |
|---|---|---|---|---|---|
|   |   | 1 | 2 | 3 | 4 |
| P | 1 | 31, 45, 46, 43 | 82, 110, 88, 72 | 43, 45, 63, 76 | 45, 71, 66, 62 |
| I | 2 | 36, 29, 40, 23 | 92, 61, 49, 124 | 44, 35, 31, 40 | 56, 102, 71, 38 |
| N | 3 | 22, 21, 18, 23 | 30, 37, 38, 29 | 23, 25, 24, 22 | 30, 36, 31, 33 |

The mean survival times of the four treatment subgroups are: 31.4, 67.6, 39.2, and 53.4 days, respectively.

The mean survival times of the three poison subgroups are: 61.8, 54.4, and 27.6 days, respectively.

The grand mean is 47.9.

This experiment is balanced, and our first step is to check for interactions; we subtract the treatment and poison means and add the grand mean to each observation to obtain the table of residuals, Table 8.8.

The statistic $I = 3.5$; a value this extreme occurs in less than 3% of the rearrangements, suggesting that the effects of treatment and poison are not additive.[9]

Table 8.8. Residuals After Subtracting Treatment and Poison Mean Effects

|   |   | Treatment | | | |
|---|---|---|---|---|---|
|   |   | 1 | 2 | 3 | 4 |
| P | 1 | $-14, 0, 1, 2$ | $1, 30, 8, -9$ | $-10, -8, 10, 23$ | $-22, 5, -1, -5$ |
| I | 2 | $-2, -9, 2, -15$ | $18, -13, -25, 46$ | $-2, -11, -15, -6$ | $-4, 42, 11, -22$ |
| N | 3 | $11, 10, 7, 12$ | $-17, -10, -9, -18$ | $4, 6, 5, 3$ | $-3, +3, -2, 0$ |

We *cannot* test to see if the main effects are significant as they are confounded with the interactions. We can test to see whether there are significant differences among treatments when our attention is restricted to one of the poisons as in the following table.

|   | Treatment | | | |
|---|---|---|---|---|
|   | 1 | 2 | 3 | 4 |
| 1 | 31, 45, 46, 43 | 82, 110, 88, 72 | 43, 45, 63, 76 | 45, 71, 66, 62 |

The treatment means are 41.25, 88.0, 56.75, 61.0; the grand mean is 61.75. The statistic $F_1$, the sum of the absolute deviations about the grand mean, is 52.5. The probability a value this large will occur purely by chance is less than 1%. We conclude there are significant differences among treatments when poison 1 is used.

## 8.4    Designing an Experiment

All the preceding results are based on the assumption that the assignment of treatments to plots (or subjects) is made at random. While it might be convenient to fertilize our plots as shown in Table 8.9, the result could be a systematic bias, particularly if, for example, there is a gradient in dissolved minerals from east to west across the field.

---

[9]By contrast, the standard parametric (ANOVA) approach yields a $p$ value greater than 10%.

Table 8.9. Systematic Assignment of Fertilizer Levels to Plots

| Hi | Med | Lo |
|----|-----|----|
| Hi | Med | Lo |
| Hi | Med | Lo |

Table 8.10. Random Assignment of Fertilizer Levels to Plots

| Hi | Med | Lo  |
|----|-----|-----|
| Lo | Med | Lo  |
| Hi | Hi  | Med |

The layout adopted in Table 8.10, obtained with the aid of a computerized random number generator, reduces but does not eliminate the effects of this hypothetical gradient. Because this layout was selected at random, the exchangeability of the error terms and, hence, the exactness of the corresponding permutation test is ensured. Unfortunately, the layout of Table 8.9 with its built-in bias can also result from a random assignment; its selection is neither more nor less probable than any of the other $\binom{9}{3\ 3}$ possibilities.

What can we do to avoid such an undesirable event? In the layout of Table 8.11, known as a Latin Square, each fertilizer level occurs once and once only in each row and in each column; if there is a systematic gradient of minerals in the soil, then this layout ensures the gradient will have almost equal impact on each of the three treatment levels. It will have an almost equal impact, even if the gradient extends from northeast to southwest rather than from east to west or north to south. I use the phrase "almost equal" because a gradient effect may still persist. The design and analysis of Latin Squares is described in the next section.

## 8.4.1   Latin Square

The Latin Square is one of the simplest examples of an experimental design in which the statistician takes advantage of some aspect of the model to reduce the overall sample size. A Latin Square is a three-factor experiment in which each combination of factors occurs once and once only. We can use a Latin Square as in Table 8.11 to assess the effects of soil composition on crop yield.

In this diagram, Factor 1: gypsum concentration, say, is increasing from left to right, Factor 2 is increasing from top to bottom (or from North to South), and

Table 8.11. Latin Square Assignment of Fertilizer Levels to Plots

| Hi  | Med | Lo  |
|-----|-----|-----|
| Lo  | Hi  | Med |
| Med | Lo  | Hi  |

the third factor occurs in combination with the other two in such a way that each combination of factors—row, column, and treatment, occurs once and once only.

Because of this latter restriction, there are only 12 different ways in which we can assign the varying factor levels to form a $3 \times 3$ Latin Square. Among the other 11 designs are the following where the varying levels of the third factor are denoted by the capital letters A, B, and C,

|   | 1 | 2 | 3 |
|---|---|---|---|
| 1 | A | C | B |
| 2 | B | A | C |
| 3 | C | B | A |

and

|   | 1 | 2 | 3 |
|---|---|---|---|
| 1 | C | B | A |
| 2 | B | A | C |
| 3 | A | C | B |

Assume we begin our experiment by selecting one of these 12 designs at random and planting our seeds in accordance with the indicated conditions.

Because there is only a single replication of each factor combination in a Latin Square, we cannot estimate the interactions. The Latin Square is appropriate only if we feel confident in assuming the effects of the various factors are completely additive, that is, the interaction terms are zero.

Our model for the Latin Square is

$$X_{ijk} = s_i + r_j + f_k + \epsilon_{ijk},$$

where, as always in a permutation analysis, we assume the labels on the errors $\{\epsilon_{ijk}\}$ are exchangeable. Our null hypothesis is $H: s_1 = s_2 = s_3$. If we assume an ordered alternative, $K: s_1 > s_2 > s_3$, our test statistic for the main effect is similar to the correlation statistic (Equation 8.5):

$$R = \sum_{i=1}^{3}(i - 1)(\overline{X}_{i..} - \overline{X}_{...})$$

or, equivalently, after eliminating the grand mean $\overline{X}_{...}$ which is invariant under permutations and multiplying by $n$, $R' = -X_{A..} + X_{C..}$.

We evaluate this test statistic both for the observed design and for each of the 12 possible Latin Square designs that might have been employed in this particular experiment. We reject the hypothesis of no treatment effect only if the test statistic $R$ for the original observations is an extreme value.

For example, suppose we employed Design 1 and observe

| 21 | 28 | 17 |
|----|----|----|
| 14 | 27 | 19 |
| 13 | 18 | 23 |

Then $X_{A..} = 58$, $X_{C..} = 57$ and our test statistic $R' = -1$. Had we employed Design 2, then $X_{A..} = 71$, $X_{C..} = 65$ and our test statistic $R' = -6$. With Design 3, $X_{A..} = 57$, $X_{C..} = 58$ and our test statistic is $+1$.

---

**Designing and Analyzing a Latin Square**

$H$: mean/median the same for all levels of each treatment.
$K$: means/medians are different for at least one level.
Assumptions:

1. observations are exchangeable if the hypothesis is true.

2. treatment effects are additive (not synergistic or antagonistic).

Procedure:
List all possible Latin Squares for the given number
of treatment levels.
Assign one at random and use it to perform the experiment.
Choose a test statistic ($R$ or $F_1$).
Compute the statistic for the design you used.
Compute the test statistic for all the other possible designs.
Determine whether the original value of the test statistic is
an extreme one.

---

We see from the permutation distribution obtained in this manner that $-1$, the value of our test statistic for the design actually employed in the experiment, is an average value, not an extreme one. We accept the null hypothesis and conclude that increasing the treatment level from A to B to C does not significantly increase the yield.

## 8.5   Determining Sample Size

Power is an increasing function of sample size, as we saw in Figures 6.2 and 6.3, so we should take as large a sample as we can afford, providing the gain in power from each new observation is worth the expense of gathering it.

For one- and two-sample comparisons of means and proportions, a number of commercially available statistics packages can help us with our determination. We've included a sample calculation with Stata in a sidebar. Note that to run Stata or similar programs, we need to specify in advance the alternative of interest and the desired significance level.

To determine sample size for more complicated experimental designs, we need to run a computer simulation. Either we specify the underlying distribution on theoretical grounds—as a normal or mixture of normals, a gamma, a Weibull, or some other well-tabulated function—or we utilize the empirical distribution obtained in some earlier experiment.

We also need to specify the significance level, e.g., 5%, and the alternative of interest, e.g., values in the second sample average 2 points larger than in the first.

To simulate the sampling process for a known distribution like the normal, we first choose a uniformly distributed random number between 0 and 1, the same range of values taken by the distribution function, then we look this number up as an entry in a table of the inverse of the normal distribution.

Programming in Stata, for example, we write **invnorm(uniform())** and repeat this command for each element in the untreated sample. If our alternative is that the population comes from a population with mean 5 and standard deviation 2, we would write 5 **+2∗invnorm(uniform())**. We repeat this command for each observation in the treated sample.

---

**Using Stata to Determine Sample Size**

Study of effect of oral contraceptives (OC) on blood pressure of women ages 35–39.
OC Users $133 \pm 15$

NonUsers $127 \pm 18$

**. sampsi 133 127, alpha(0.05) power(0.8) sd1(15) sd2(18)**

Estimated sample size for two-sample comparison of means

Test Ho: $m_1 = m_2$, where $m_1$ is the mean in population 1 and $m_2$ is the mean in population 2
Assumptions:

$$\text{alpha} = 0.0500(\text{two-sided})$$
$$\text{power} = 0.8000$$
$$m_1 = 133$$
$$m_2 = 127$$
$$sd_1 = 15$$
$$sd_2 = 18$$
$$n_2/n_1 = 1.00$$

Estimated required sample sizes:

$$n_1 = 120$$
$$n_2 = 120$$

---

If we aren't sure about the underlying distribution, we draw a bootstrap sample with replacement from the empirical distribution. We compute the test statistic for the sample, and note whether we accept or reject. We repeat the entire process 50–400 times (50 times when just trying to get a rough idea of the correct sample size, 400 times when closing in on the final value). The number of rejections divided by the number of simulations provides us with an estimate of the power for the specific experimental design and our initial sample sizes. If the power is still too low, we increase the sample sizes and repeat the preceding simulation process.

## 8.6    Unbalanced Designs

Imbalance in the design will result in the *confounding* of main effects with interactions. Consider the following two-factor model for crop yield:

$$X_{ijk} = \mu + s_i + r_j + (sr)_{ij} + \epsilon_{ijk}.$$

Now, suppose the observations in a two-factor experimental design are distributed as in the following diagram:

| Mean 0 | Mean 2 |
|---|---|
| Mean 2 | Mean 0 |

There are no main effects in this example—both row means and both column means have the same expectation, but there is a clear interaction represented by the two nonzero off-diagonal elements.

If the design is balanced, with equal numbers per cell, the lack of significant main effects and the presence of a significant interaction should and will be confirmed by our analysis. But suppose the design is not in balance, that for every 10 observations in the first column, we have only one observation in the second. Because of this imbalance, when we use the statistic $S$ (Equation 8.1), we will uncover a false "row" effect that is actually due to the interaction between rows and columns. The main effect is said to be *confounded* with the interaction.

If a design is unbalanced as in the preceding example, we cannot test for a "pure" main effect or a "pure" interaction. But we may be able to test for the combination of a main effect with an interaction by using the statistic ($S$, $F_1$, or $R$) that we would use to test for the main effect alone. This combined effect will not be confounded with the main effects of other unrelated factors.

### 8.6.1    Multidimensional Contingency Tables

Whether or not the design is balanced, the methods of Chapter 7 can be applied to multifactor designs whose outcomes are either categorical or of the yes/no variety. Table 8.12 contains the results of an experiment by Plackett and Hewlett [1963] in which milkweed bugs were exposed to various levels of two insecticides. At issue is whether the two drugs act independently.

Table 8.12. Deaths of Milkweed Bugs Exposed to Various Levels of Two Insecticides

| | DoseB = 0 | | | DoseB = 0.2 | | |
|---|---|---|---|---|---|---|
| Dose A | 0 | 0.05 | 0.07 | 0 | 0.05 | 0.07 |
| Died | | 9 | 22 | 5 | 27 | 27 |
| Survived | | 39 | 26 | 43 | 21 | 21 |

Although death due to a variety of spontaneous and background causes could be anticipated, no attempt was made to actually measure this background—the cell corresponding to a zero dose of each drug is empty. The resultant design is an unbalanced one. Still, two solutions are possible via either the permutation method or the bootstrap.

## Permutation

Once again we turn to StatXact, this time using the Stratified 2 x C Tables option. We analyze the data separately for each level of Insecticide B using the Cochran-Armitage test for trend to obtain estimates of the dose effects of Insecticide A. The results are

Dose $B = 0$    Trend is 1.29 with a 95% confidence interval of $(0.3, 2.3)$

Dose $B = 0.2$ Trend is 1.02 with a 95% confidence interval of $(0.6, 1.5)$.

As the confidence intervals are overlapping, we have no reason to believe the effects of the two insecticides are not additive.

## Bootstrap

An underlying biological assumption is that the dose threshold above which a given insecticide is toxic varies from insect to insect. Suppose we form a pair of bootstrap samples as in Wahrenlot [1980]. The first sample we construct in two stages: first, we draw an observation at random from the sample of 48 milkweed bugs treated with 0.05 units of the first insecticide alone. If by chance we select one of the 39 survivors, then we draw from the sample of 48 bugs treated with 0.2 units of the second insecticide alone. Otherwise, we record a "death."

Of course, we don't actually perform the drawing but simulate it through the use of a random number generator. If this number is greater than 9/48, the insect lives to be treated a second time, otherwise it dies.

The second bootstrap sample we select with replacement from the 27 killed and 21 survivors in the sample treated with both insecticides. We repeat the process 50–200 times, each time comparing the number of survivors in the two bootstrap samples. If the two insecticides act independently, the numbers should be comparable. See, also, Romano [1988].

## 8.6.2   *Missing Combinations*

If an entire factor-combination is missing, we may not be able to estimate or test any of the effects. One very concrete example is an unbalanced design I encountered in the 1970s when I worked with Makinodan et al. [1976] to study the effects of age on the mediation of the immune response. They measured the anti-SBRC response of spleen cells derived from C57BL mice of various ages. In one set of trials, the cells were derived entirely from the spleens of young mice, in a second, they came from the spleens of old mice, and in a third they came from mixtures of the two.

Let $X_{ijk}$ denote the response of the kth sample taken from a population of type $i$, $j$, where $i = 0 = j$ denotes controls; $i = 1$, $j = 0$ denotes cells from young animals only; $i = 0$, $j = 1$ denotes cells from old animals only; and $i = 1 = j$ denotes a mixture of cells from old and young animals. We assume for lymphocytes taken from the spleens of young animals,

$$X_{10k} = \mu + \alpha + \epsilon_{10k},$$

for the spleens of old animals,

$$X_{01k} = \mu - \alpha + \epsilon_{01k},$$

and for a mixture of $p$ spleens from young animals and $(1 - p)$ spleens from old animals, where $0 \le p \le 1$,

$$X_{11k} = p(\mu + \alpha) + (1 - p)(\mu - \alpha) - \gamma + \epsilon_{11k}$$
$$= \mu + (1 - 2p)(\alpha) - \gamma + \epsilon_{11k}$$

where the $\{\epsilon_{11k}\}$ are independent values.

Makinodan knew in advance of his experiment that the main effect $\alpha$ was non-zero. He also knew the distributions of the errors $\epsilon_{11k}$ would be different for the different populations. We can assume only that these errors are independent of one another and that their medians are zero.

Makinodan wanted to test the hypothesis that the interaction term $\gamma > 0$ as there are immediate biological interpretations for the three alternatives: from $\gamma = 0$ one may infer independent action of the two cell populations; $\gamma < 0$ means excess lymphocytes in young populations; and $\gamma > 0$ suggests the presence of suppressor cells in the spleens of older animals.

But what statistic are we to use to do the test? One possibility is

$$S = |\overline{X}_{11.} - p\overline{X}_{01.} - (1 - p)\overline{X}_{10.}|.$$

If the design were balanced, or we could be sure that the null effect $\mu = 0$, this is the statistic we would use. But the design is not balanced, with the result the main effects with which we are not interested are confounded with the interaction with which we are.[10]

---

[10]The standard parametric (ANOVA) approach won't work in this example either.

---

**Mean DPFC Response**

Effect of pooled old BC3FL spleen cells on the anti-SRBC response of indicator pooled BC3FL spleen cells. Data extracted from Makinodan et al. [1976]. Bootstrap analysis.

| Young Cells | Old Cells | 1/2 + 1/2 |
|---|---|---|
| 5640 | 1150 | 7100 |
| 5120 | 2520 | 11020 |
| 780 | 900 | 13065 |
| 4430 | 50 | |
| 7230 | | |

Bootstrap sample 1:   $5640 + 900 - 11020 - 4480$
Bootstrap sample 2:   $5780 + 1150 - 11020 - 4090$
Bootstrap sample 3:   $7230 + 1150 - 7100 + 1280$

. . . . . . . . . . . . . . .
. . . . . . . . . . . . . .

Bootstrap sample 100:   $5780 + 2520 - 7100 + 1200$

---

Fortunately, another resampling method, the bootstrap, can provide a solution: Draw an observation at random and with replacement from the set $\{x_{10k}\}$; label it $x_{10}^*$. Similarly, draw the bootstrap observations $x_{01}^*$ and $x_{11}^*$ from the sets $\{x_{01k}\}$ and $\{x_{11k}\}$. Let

$$\gamma^* = px_{01}^* + (1 - p)x_{10}^* - x_{11}^*.$$

Repeat this resampling procedure several hundred times, obtaining a bootstrap estimate $\gamma^*$ of the interaction each time you resample. Use the resultant set of bootstrap estimates $\{\gamma^*\}$ to obtain a confidence interval for $\gamma$. If 0 belongs to this confidence interval, accept the hypothesis of additivity; otherwise reject. Our results are depicted in Figure 8.2.

The sample size in this experiment is small, and if we were to rely on this sample alone we could not rule out the possibility that $\gamma \geq 0$. Indeed, the 90% confidence interval stretches from $-7815$ to $+1060$. But Makinodan et al. [1976] conducted many replications of this experiment for varying values of $p$, with comparable results; they could feel confident in concluding $\gamma < 0$ showing that young spleens have an excess of lymphocytes.

## 8.7   Summary

In this chapter, you learned the principles of experimental design: to block or measure all factors under your control, to randomize with regard to factors that are not.

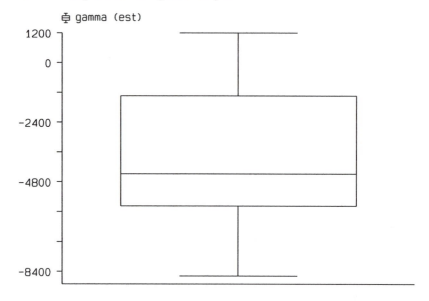

**Figure 8.2.**    Box and whiskers plot of the bootstrap distribution of $\gamma^*$.

You learned to analyze balanced $k$-way designs for main effects, and balanced two-way designs for both main effects and interactions. You learned to use the Latin Square to reduce sample size and to use bootstrap methods when designs are not balanced.

## 8.8    To Learn More

For more on the principles of experimental design, see Fisher [1935], Kempthorne [1955], Wilk and Kempthorne [1956, 1957], Scheffe [1959], Maxwell and Cole [1991]. Further sample-size guidelines are provided in Shuster [1993].

Permutation tests have been applied to a wide variety of experimental designs including analysis of variance [Kempthorne 1955, 1966, 1977; Kempthorne and Doerfler, 1969; Jin 1984; Soms 1985; Diggle, Lange, and Benes 1991; Ryan, Tracey, and Rounds 1996, Pesarin 2001], clinical trials [Lachin 1988a,b], covariance [Peritz, 1982], crossovers [Shen and Quade 1986], factorial [Loughin and Noble 1997], growth curves [Foutz, Jensen, and Anderson 1985; Zerbe 1979a,b], matched pairs [Peritz 1985; Welch 1987; Welch and Gutierrez, 1988; Rosenbaum 1988; Good 1991, Zumbo 1996], randomized blocks [Wilk 1955], restricted randomization [Smythe 1988], sequential clinical trials [Wei 1988; Wei, Smythe and Smith 1986]. Bradbury [1987] compares parametric with randomization tests. Mapleson [1986] applied the bootstrap to the analysis of clinical trials.

## 8.9   Exercises

1. Design an experiment.
   a. List all the factors that might influence the outcome of your experiment.
   b. Write a model in terms of these factors.
   c. Which factors are under your control?
   d. Which of these factors will you use to restrict the scope of the experiment?
   e. Which of these factors will you use to block?
   f. Which of the remaining factors will you neglect initially, that is, lump into your error term?
   g. How will you deal with each of the remaining covariates?
   h. How many subjects/items will you observe in each subcategory?
   i. Write out two of the possible assignments of subjects to treatment.
   j. How many possible assignments are there in all?

2. A known standard was sent to six laboratories for testing.
   a. Are the results comparable among laboratories?

   |      | laboratory | | | | | |
   |------|-------|-------|-------|-------|-------|-------|
   |      | A     | B     | C     | D     | E     | F     |
   | 1/1  | 221.1 | 208.8 | 211.1 | 208.3 | 221.1 | 224.2 |
   | 1/2  | 224.2 | 206.9 | 198.4 | 214.1 | 208.8 | 206.9 |
   | 1/3  | 217.8 | 205.9 | 213.0 | 209.1 | 211.1 | 198.4 |

   b. The standard was submitted to the same laboratories the following month. Are the results comparable from month to month?

   |      | laboratory | | | | | |
   |------|-------|-------|-------|-------|-------|-------|
   |      | A     | B     | C     | D     | E     | F     |
   | 2/1  | 208.8 | 211.4 | 208.9 | 207.7 | 208.3 | 214.1 |
   | 2/2  | 212.6 | 205.8 | 206.0 | 216.2 | 208.8 | 212.6 |
   | 2/3  | 213.3 | 202.5 | 209.8 | 203.7 | 211.4 | 205.8 |

3. Potted tomato plants in groups of six were maintained in one of three levels of light and two levels of water. What effects, if any, did the different levels have on yield?

   | Light | Water | Yield | Light | Water | Yield | Light | Water | Yield |
   |-------|-------|-------|-------|-------|-------|-------|-------|-------|
   | 1     | 1     | 12    | 2     | 2     | 16    | 3     | 1     | 17    |
   | 1     | 1     | 8     | 2     | 2     | 12    | 3     | 1     | 24    |
   | 1     | 1     | 8     | 2     | 2     | 13    | 3     | 1     | 22    |
   | 1     | 2     | 13    | 2     | 2     | 19    | 3     | 2     | 24    |
   | 1     | 2     | 15    | 2     | 2     | 16    | 3     | 2     | 26    |
   | 1     | 2     | 16    | 2     | 2     | 17    | 3     | 2     | 28    |

**4. a.** Are the two methods of randomization described in 8.1.3 equivalent?

   **b.** Suppose you had a six-sided die and three coins. How would you assign plots to one of four treatments? three rows and two columns? eight treatments?

**5.** Do the four hospitals considered in Exercise 16 of Chapter 1 have the same billing practices? (Hint: Block the data before analyzing.)

**6.** Four containers, each with 10 oysters, were randomly assigned to four stations each kept at a different temperature in the wastewater canal of a power plant. The containers were weighed before treatment and after one month in the water. Were there statistically significant differences among the stations?

| Trt | Initial | Final | Trt | Initial | Final |
|-----|---------|-------|-----|---------|-------|
| 1 | 27.2 | 32.6 | 3 | 28.9 | 33.8 |
| 1 | 31.0 | 35.6 | 3 | 23.1 | 29.2 |
| 1 | 32.0 | 35.6 | 3 | 24.4 | 27.6 |
| 1 | 27.8 | 30.8 | 3 | 25.0 | 30.8 |
| 2 | 29.5 | 33.7 | 4 | 29.3 | 34.8 |
| 2 | 27.8 | 31.3 | 4 | 30.2 | 36.5 |
| 2 | 26.3 | 30.4 | 4 | 25.5 | 30.8 |
| 2 | 27.0 | 31.0 | 4 | 22.7 | 25.9 |

**7.** You can increase the power of a statistical test in three ways: a) making additional observations, b) making more precise observations, c) adding covariates. Discuss this remark in the light of your own experimental efforts.

**8.** A pregnant animal was accidentally exposed to a high dose of radioactivity. Her seven surviving offspring (one died at birth) were examined for defects. Three tissue samples were taken from each animal and examined by a pathologist. What is the sample size?

**9.** Show that if $T'$ is a monotonic increasing function of $T$, that is, $T'$ always increases when $T$ increases, then a test based on the permutation distribution of $T'$ will accept or reject only if a permutation test based on $T$ also accepts or rejects.

**10.** Should your tax money be used to fund public television? Recall from Exercise 11 when a random sample of adults were asked for their responses on a 9-point scale (1 is very favorable and 9 is totally opposed) the results were as follows:

$$3, 4, 6, 2, 1, 1, 5, 7, 4, 3, 8, 7, 6, 9, 5.$$

The first 10 of these responses came from the City, and the last five came from the Burbs. Are there significant differences between the two areas? Given that two-thirds of the population live in the City, provide a point estimate and confidence interval for the mean response of the entire population. [Hint: modify the test statistic so as to weight each block proportionately.]

11. Take a third look at the experimental approaches described in Exercises 12 and 13 of Chapter 3. What might be the possible drawbacks of each? How would you improve on the methodology?

12. Your company is considering making a take-over bid for a marginally profitable line of retail stores. Some of the stores are real moneymakers; others could be a major drain on your own company's balance sheet. You still haven't made up your mind, when you get hold of the following figures based on a sample of 15 of the stores:

| Store Size | Percent Profit |
|---|---|
| small | 7.0, 7.3, 6.5, 6.4, 7.5 |
| medium | 7.3, 8.0, 8.5, 8.2, 6.8 |
| large | 8.7, 8.1, 8.9, 9.2, 9.5 |

Have these figures helped you make up your mind? Write up a brief report for your CEO summarizing your analysis.

13. Using the insecticide data collected by Plackett and Hewlett [1963] in Table 8.6, test for independence of action using only the zero and highest dose level of the first drug. Can you devise a single test that would utilize the data from all dose levels simultaneously?

14. Aflatoxin is a common and undesirable contaminant of peanut butter. Are there significant differences among the following brands?

| Snoopy | 0.5 | 7.3 | 1.1 | 2.7 | 5.5 | 4.3 |
|---|---|---|---|---|---|---|
| Quick | 2.5 | 1.8 | 3.6 | 5.2 | 1.2 | 0.7 |
| Mrs. Good's | 3.3 | 1.5 | 0.4 | 4.8 | 2.2 | 1.0 |

The aflatoxin content is given in ppb.

15. In Exercise 5.9, we considered the effect of noise on productivity. The following table includes additional information concerning years of experience. Presumably, more experienced workers should be less bothered by noise. But is this the case?

| Noise db | Duration min | Experience yrs | Noise db | Duration min | Experience yrs |
|---|---|---|---|---|---|
| 0 | 11.5 | 4 | 40 | 16.0 | 8 |
| 0 | 9.5 | 7 | 60 | 20.0 | 5 |
| 20 | 12.5 | 3 | 60 | 19.5 | 7 |
| 20 | 13.0 | 2 | 80 | 28.5 | 8 |
| 40 | 15.0 | 11 | 80 | 27.5 | 10 |

**16.** In how many different ways can we assign 9 subjects to 3 treatments, given equal numbers in each sample? What if we began with 6 men and 3 women and wanted to block our experiment?

**17.** Unequal sample sizes? Take a second look at the data of Section 8.6.2. We seem to have three different samples with three different samples sizes, each drawn from a continuous domain of possible values. Or do we? From the viewpoint of the bootstrap, each sample represents our best guestimate of the composition of the larger population from which it was drawn. Each hypothetical population appears to consist of only a finite number of distinct values, a different number for each of the different populations. Discuss this seeming paradox.

# CHAPTER 9

# Multiple Variables and Multiple Hypotheses

The value of an analysis based on simultaneous observations on several variable such as height, weight, blood pressure, and cholesterol level is that it can be used to detect subtle changes that might not be detectable, except with very large, prohibitively expensive samples, were we to consider only one variable at a time.

Any of the resampling procedures can be applied in a multivariate setting providing we can either

a) find a single-valued test statistic that can stand in place of the multivalued vector of observations

b) combine the $p$-values associated with the various univariate tests into a single $p$-value.

Let us consider each of these approaches in turn along with the generalized quadratic form and the challenging problems of repeated measures and multiple hypotheses.

---

**Increase the Sensitivity of Your Experiments**

- Take more observations
- Take more precise observations
- Block your experiments
- Observe multiple variables

---

## 9.1   Single-Valued Test Statistic

Suppose a small college has found that the best predictor of academic success is a weighted combination of SAT scores, adjusted GPA, and the number of extra-curricular activities. Then, it would use this same weighted combination when testing hypotheses concerning, say, the relative merits of in-state and out-of-state students.

In the abscense of such domain specific knowledge, Hotelling's $T^2$, a straightforward generalization of Student's $t$ (discussed in Section 4.3.2) may be used. Suppose we have made a series of exchangeable vector-valued observations consisting of $J$ variables each. In the $i$th vector of observations $\vec{X}_i = \{X_i^1, X_i^2, \ldots, X_i^J\}$, $X_i^1$ might denote the height of the $i$th subject, $X_i^2$ his weight, and so forth. Let $\vec{X}$. denote the vector of mean values $\{X^1, X^2, \ldots, X^J\}$ and $V$ the $J \times J$ covariance matrix whose $ij$th element $V_{ij}$ is the covariance of $X^i$ and $X^j$ (see Section 5.3). To test the hypothesis that the midvalue of $\vec{X}_i$ is $\vec{\xi}_i = \{\xi_i^1, \xi_i^2, \ldots, \xi_i^J\}$, use Hotelling's $T^2 = (\vec{X}. - \vec{\xi})V^{-1}(\vec{X}. - \vec{\xi}).$[1]

This statistic weighs the contribution of individual variables and pairs of variables in inverse proportion to their covariances. This has the effect of rescaling each variable so that the most weight is given to those variables that can be measured with the greatest precision and those that can provide information not provided by the others.

For the purpose of resampling, each vector of observations on an individual subject is treated as a single indivisible entity. When we relabel, we relabel on a subject-by-subject basis so that all observations on a single subject receive the same new label. If the original vector of observations on subject $i$ consists of $j$ distinct observations on $j$ different variables $(x_i^1, x_i^2, \ldots, x_i^j)$ and we give this vector a new label $\pi(i)$, then the individual observations remain together as a unit, each with the same new label $(x_{\pi(i)}^1, x_{\pi(i)}^2, \ldots, x_{\pi(i)}^j)$.

## 9.1.1    Two-Sample Multivariate Comparison

The two-sample multivariate comparison is only slightly more complicated. Let $n_1$, $\vec{X}_{1.}$; $n_2$, $\vec{X}_{2.}$ denote the sample size and vector of mean values of the first and second samples, respectively. Under the null hypothesis, we assume that the two sets of vector-valued observations come from the same distribution (that is, the labels 1 and 2 are exchangeable between the samples). Let $V$ denote the pooled estimate of the common covariance matrix; that is, $V_{ij}$ is the covariance estimate of $X^i$ and $X^j$ based on the combined sample.

$$(N - 2)V_{ij} = \sum_{m=1}^{2} \sum_{k=1}^{n_m} (X_{mk}^i - X_{m.}^i)(X_{mk}^j - X_{m.}^j).$$

To test the hypothesis that the midvalues of the two distributions are the same, we could use the statistic

$$T^2 = (\vec{X}_{1.} - \vec{X}_{2.})V^{-1}(\vec{X}_{1.} - \vec{X}_{2.})^T,$$

---

[1] A definition of $V^{-1}$ may be found in texts on matrix or vector algebra, but why reinvent the wheel? Routines for manipulating and, in this case, inverting matrices are readily available in Gauss, IMSL, SAS, S, and a number of C libraries.

but then we would be forced to recompute the covariance matrix $V$ and its inverse $V^{-1}$ for each new rearrangement. To reduce the number of computations, Wald and Wolfowitz [1943] suggest a slightly different statistic $T'$ that is a monotonic[2] function of $T$.

$$\text{Let } U_j = \frac{1}{N} \sum_{i=1}^{2} \sum_{k=1}^{n_i} X_{ikj},$$

$$c_{ij} = \sum_{m=1}^{2} \sum_{k=1}^{n_i} (X_{mki} - U_i)(X_{mkj} - U_j).$$

Let $C$ be the matrix whose components are the $c_{ij}$. Then

$$T'^2 = (\vec{X}_{1.} - \vec{X}_{2.})C^{-1}(\vec{X}_{1.} - \vec{X}_{2.})^T.$$

---

**Calculating the Wald-Wolfowitz Variant of Hotelling's $T^2$**
**Blood Chemistry Data from Warner et al. [1990]**

| ID | BC | Albumin | Uric Acid | Mean | Albumin | Uric Acid |
|---|---|---|---|---|---|---|
| 2381 | N | 43 | 54 | N | 41.25 | 47.25 |
| 1610 | N | 41 | 33 | Y | 37.0 | 52.75 |
| 1149 | N | 39 | 50 | Y–N | −4.25 | 7.50 |
| 2271 | N | 42 | 48 | | | |
| 1946 | Y | 35 | 72 | C | | |
| 1797 | Y | 38 | 30 | | 8.982 | −21.071 |
| 575 | Y | 40 | 46 | | −21.071 | 197.571 |
| 39 | Y | 35 | 63 | | | |
| | | | | $C^{-1}$ | | |
| | | | | | 0.1487 | 0.01594 |
| | | | | | 0.01594 | 0.006796 |

Hotelling's $T^2$
$= (-4.25 \; 7.50)C^{-1}(-4.25 \; 7.50)^T = 2.092$

---

As with all permutation tests we proceed in three steps:

1. We compute the test statistic for the original observations,

2. We compute the test statistic for all relabelings,

3. We determine the percentage of relabelings that lead to values of the test statistic that are as or more extreme than the original value.

---

[2]A monotonic increasing function increases whenever its argument increases. An example of a monotonic increasing function is $f[x] = 32 + 9.2x$, another is $f[x] = \exp[x]$. A monotonic decreasing function decreases whenever its argument increases. If $x > 0$, then $x^2$ is a monotonic increasing function of $x$; but $x^2$ is not a monotonic function for all $x$ as it decreases for $x < 0$ until it reaches a minimum at 0 and then increases for $x > 0$.

Hotelling's $T^2$ is the appropriate statistic to use if you suspect the data have a distribution close to that of the multivariate normal. Under the assumption of multivariate normality, the power of the permutation version of Hotelling's $T^2$ converges with increasing sample size to the power of the most powerful parametric test that is invariant under transformations of scale. As is the case with its univariate counterpart Student's $t$, Hotelling's $T^2$ is most powerful among unbiased tests for testing a general hypothesis that all the observations are drawn from the same multivariate distribution against a specific normal alternative (Runger and Eaton, 1992).

While the stated significance level of the parametric version of Hotelling's $T^2$ cannot be relied on for small samples if the data is not normally distributed (Davis, 1982), as always, the corresponding permutation test yields an exact significance level, providing the errors are exchangeable from sample to sample.

### 9.1.2  Applications to Repeated Measures

When we do a bioequivalence study, we replace a set of discrete values with a "smooth" curve. This curve is derived in one of two ways: 1) by numerical analysis, 2) by modeling. The first way yields a set of coefficients, the second a set of parameter estimates. Either the coefficients or the estimates may be treated as if they were the components of a multivariate vector and the methods of the previous section applied to them.

Here is an elementary example: Suppose you observe the time course of a drug in the urine over a period for which a linear model would be appropriate. Suppose further that the chief virtue of your measuring system is its low cost so that measurement errors are significant. Consequently, you take a series of measurements on each patient about half an hour apart and then use least squares methods[3] to derive a best-fitting line for each patient. That is, following an approach first suggested by Zerbe and Walker [1977], you replace the set of measurements $\{X_{ijk}\}$ where $i = 0$ or $1$ denotes the drug used, $j = 1, \ldots, J$ denotes the subject, and $k = 1, \ldots, K$ denotes a specific observation, with the set of vectors $\{\vec{Y}_{ij} = (a_{ij}, b_{ij})\}$ where $a_{ij}$ and $b_{ij}$ are the intercept and slope of the regression line for the $j$th subject in the $i$th treatment group.

Using the methods described in the preceding section, you calculate the mean vector and the covariance matrix for the $\{\vec{Y}_{ij}\}$, and compute Hotelling's $T^2$ for the original observations and for a set of random arrangements. Then use the resultant permutation distribution to determine whether the time courses of the two drugs are similar.

The preceding is just one of the many possible experiments in which we study the development of a process over a period of time, such as the growth of a tumor or the gradual progress of a cure. If our observations are made by sacrificing different groups of animals at different periods of time, then time is simply another variable in the analysis which we may treat as a covariate using the methods of

---

[3]See Chapter 10.

Chapters 8 and 10. But if all our observations are made on the same subjects, then the multiple observations on a single individual will be interdependent. And all the observations on a single subject must be treated as a single multivariate vector.

We may ask at least three questions about the response profiles:

1. Are the response profiles the same for the various treatments?

2. Are the response profiles parallel?

3. Are the response profiles at the same level?

A "yes" answer to question 1 implies "yes" answers to questions 2 and 3, but we may get a "yes" answer to 2 even when the answer to 3 is "no."

One simple test of parallelism entails computing the successive differences $z_{j,i} = x_{j,i+1} - x_{j,i}$ for $j = 1, 2; i = 1, \ldots, I - 1$ and then applying the methods of the preceding section to these differences. Of course, this approach is applicable only if the observations on both treatments were made at identical times.

To circumvent this limitation and to obtain a test of the narrower hypothesis of equivalent response profiles, Koziol et al. [1981] use an approach based on ranks:

Suppose there are $N_i$ subjects in group $i$. Let $X^i_{tj}$, $t = 1, 2, \ldots, T$; $j = 1, 2, \ldots, N_i$ denote the observation on the $j$th subject in Group $i$ at time $t$. Not all the $X^i_{tj}$ may be observed in practice; we may only have observations for $N_{it}$ of the $N_i$ in the $i$th group at time $t$. Let $R^i_{tj}$ be the rank of $X^i_{tj}$ among these $N_{it}$ values. Let $S_{it}$ denote the mean of these ranks.

If luck is with us so that all subjects remain with us to the end of the experiment, then $N_{it} = N_i$ for all $t$ and each $i$, and we may adopt a test statistic first proposed by Puri and Sen [1966]

$$L = \sum_i N_i \vec{S}_i V^{-1} \vec{S}_i^T,$$

where $\vec{S}_i$ is a $1 \times T$ vector with components $(S_{it}, S_{it}, \ldots, S_{it})$ and $V$ is a $T \times T$ covariance matrix whose $st$th component is

$$v_{st} = \sum_i \sum_j R^i_{sj} R^i_{tj} / T N_i.$$

## 9.2    Combining Univariate Tests

The methods described in this section have the advantage that they apply to continuous, ordinal, or categorical variables or to any combination thereof. They can be applied to one-, two- or $k$-sample comparisons. The weakness is that the variables must be independent. As in the preceding section suppose we have made a series of exchangeable vector-valued observations on $J$ variables. In the $k$th vector of observations in the $i$th treatment group, $\vec{X}_{ik} = \{X^1_{ik}, X^2_{ik}, \ldots, X^J_{ik}\}, X^1_{ik}$ might be a 0 or 1 according to whether or not the $k$th seedling in the $i$th treatment group germinated, $X^2_{ik}$ the height of the $k$th seedling, $X^3_{ik}$ its weight, $X^4_{ik}$ a particular type of soil, and so forth. Let $\vec{T}_o$ denote the vector of single-variable test statistics $\{T_{01}, T_{02}, \ldots, T_{0J}\}$ based on the original unpermuted matrix of observations $X_o$. These might include differences of means or weighted sums of the total number germinated, or any other statistic one might employ when testing just one variable at a time. When we permute the observation vectors among treatments, we obtain a new vector of single-variable test statistics $\{T_1^*, T_2^*, \ldots, T_J^*\}$.

In order to combine the various tests, we need to reduce them all to a common scale. Proceed as follows:

1. Generate $S$ permutations of $X$ and thus obtain $S$ vectors of univariate test statistics.

2. Rank each of the single-variable test statistics separately. Let $R_{ij}$ denote the rank of the test statistic for the $j$th variable for the $i$th permutation when it is compared to the values of the test statistic for the $j$th variable for the other $S - 1$ permutations.

3. Combine the ranks of the various univariate tests for each permutation using Fisher's omnibus method:

$$U_i = -\sum_{j=1}^{J} \log \left( \frac{S + 0.5 - R_{ij}}{S + 1} \right); \qquad i = 1, \ldots, S.$$

4. Reject the null hypothesis only if the value of $U$ for the original observations is an extreme value.

This straightforward, yet powerful method is due to Pesarin [ 1990].

## 9.3    The Generalized Quadratic Form

### 9.3.1    Mantel's U

Mantel's $U = \sum \sum a_{ij} b_{ij}$ is perhaps the most widely used of all multivariate statistics because of its broad range of applications. By appropriately restricting the values of $a_{ij}$ and $b_{ij}$, the definition of Mantel's $U$ can be seen to include several of the standard measures of correlation including those usually attributed to Pearson, Pitman, Kendall, and Spearman [Hubert 1985]. In Mantel's original formulation, Mantel [1967], $a_{ij}$ is a measure of the temporal distance between items $i$ and $j$, while $b_{ij}$ is a measure of the spatial distance. As an example, suppose $t_i$ is the day on which the $i$th individual in a study came down with cholera and $(x_i, y_i)$ denotes her position in space. For all $i, j$ set $a_{ij} = 1/(t_i - t_j)$ and $b_{ij} = 1/\sqrt{(x_i - x_j)^2 + (y_i - y_j)^2}$.

A large value for $U$ would support the view that cholera spreads by contagion from one household to the next. How large is large? As always, we compare the value of $U$ for the original data with the values obtained when we fix the $i$'s but permute the $j$'s as in $U' = \sum \sum a_{ij} b_{i\pi[j]}$.

Table 9.1. Incidents of Pairs of Ancephalic Infants by Distance and Months Apart

|         | months apart | | |
|---------|------|------|------|
| km apart | < 1 | < 2 | < 4 |
| < 1 | 39 | 101 | 235 |
| < 5 | 53 | 156 | 364 |
| < 25 | 211 | 652 | 1516 |

## 9.3.2   Example in Epidemiology

An ongoing fear of many parents is that something in their environment—asbestos or radon in the walls of their house, or toxic chemicals in their air and ground-water—will affect their offspring. Table 9.1 is extracted from data collected by Siemiatycki and McDonald [1972] on congenital neural-tube defects. Eyeballing the gradiant along the diagonal of this table one might infer that births of an-cephalic infants occur in clusters. We could test this hypothesis statistically using the methods of Chapter 7 for ordered categories, but a better approach since the exact time and location of each event is known is to use Mantel's $U$. The question arises as to which measures of distance and time we should employ. Mantel [1967] reports striking differences between one analysis of epidemiologic data in which the coefficients are proportional to the differences in position and a second approach (which he recommends) to the same data in which the coefficients are proportional to the reciprocals of these differences.

Using Mantel's approach, a pair of infants born 5 kilometers and 3 months apart contribute $\frac{1}{3} * \frac{1}{5} = \frac{1}{15}$ to the statistic. Summing up the contributions from all pairs, then repeating the summing process for a series of random rearrangements, Siemiatycki and McDonald conclude the clustering of ancephalic infants is not statistically significant. You can confirm their result by entering the data into StatXact, using TableData, Settings to specify the row weights (2, 0.4, 0.08) and the column weights (2, 0.67, 0.33), then selecting Statistics, Doubly Ordered $R \times C$ Table, Linear-by-linear, and Exact from the menus to obtain a $p$-value of 0.18.

## 9.3.3   Further Generalization

Mantel's $U$ is quite general in its application. The coefficents need not correspond to space and time. In a completely disparate application in sociology [Hubert and Schultz, 1976], observers studied $k$ distinct variables in each of large number of subjects. Their object was to test a specific sociological model for the relation-ships among the variables. The $\{a_{ij}\}$ in Mantel's $U$ were elements of the $k \times k$ sample-correlation matrix while the $\{b_{ij}\}$ were elements of an idealized or theo-retical correlation matrix derived from the model. A large value of $U$ supported the model; a small value would have ruled against it.

### 9.3.4  The MRPP Statistic

The MRPP or multiresponse permutation procedure [Mielke, 1979a] has been applied to applications as diverse as the weather and the spatial distribution of archaeological artifacts. The MRPP uses the permutation distribution of between-object distances to determine whether a classification structure has a nonrandom distribution in space or time.

An example of the application of the MRPP arises in the assignment of antiquities (artifacts) to specific classes based on their spatial locations in an archaeological dig. Presumably, the kitchen tools of primitive man—woks and Cuisinarts—should be found together, just as a future archaeologist can expect to find TV, VCR, and stereo side by side in a neolithic living room.

Following Berry et al. [1980, 1983], let $\Omega = \{\omega_1, \ldots, \omega_n\}$ designate a collection of $N$ artifacts within a site; let $X_{i1} \cdots X_{ir}$ denote the $r$ coordinates for the site space for the $i$th artifact; let $S_1, \ldots, S_{g+1}$ represent an exhaustive partitioning of the $N$ artifacts into $g + 1$ disjoint classes (the $g + 1$st being reserved for not-yet-classified items); and let $n_j$ be the number of artifacts in the $j$th class.

Define the distance between the $i$th and $j$th artifacts as

$$\delta_{ij} = \frac{1}{\sqrt{\sum_{k=1}^{r}(X_{ik} - X_{jk})^2}}.$$

Define the average between-artifact distance for all artifacts within the $i$th class,

$$\zeta_i = \frac{2}{n_i(n_i - 1)} \sum_{i<j} \delta_{ij} \phi_i[\omega_i] \phi_i[\omega_j],$$

where $\phi_i[\omega]$ is an indicator function that is 1 if $\omega \in S_i$ and 0 otherwise.

The test statistic is the weighted within-class average of these distances,

$$\Delta = \frac{\sum_{i=1}^{g} n_i \zeta_i}{\sum_{i=1}^{g} n_i}.$$

The permutation distribution associated with $\Delta$ is taken over all allocations of the $N$ artifacts to the $g + 1$ classes with the same numbers of artifacts $\{n_1, \ldots, n_{g+1}\}$ assigned to each class.

Sound complicated? Perhaps. But the complications are the result of an archeologist thinking in quantitative terms about a problem in archeology. The method is as straightforward as those we discussed in Chapter 3. Choose a test statistic, compute the statistic for the original observations, calculate the permutation distribution of the test statistic you've chosen, then determine the significance level.

## 9.4  Multiple Hypotheses

One of the difficulties with clinical trials and other large-scale studies is that frequently so many variables are under investigation that one or more of them is practically guaranteed to be significant by chance alone. If we perform 20 tests at the 5% or 1/20 level, we expect at least one significant result on the average. If the variables are related (and in most large-scale medical and sociological studies the variables have complex interdependencies), the number of falsely significant results could be many times greater.

A resampling procedure outlined by Troendle [1995] allows us to work around the dependencies. Suppose we have measured $k$ variables on each subject, and are now confronted with $k$ test statistics $s_1, s_2, \ldots, s_k$. To make these statistics comparable, we need to standardize them and render them dimensionless, dividing each by its respective $L_1$ or $L_2$ norm. For example, if one variable, measured in centimeters, takes values like 144, 150, 156 and the, other measured in meters, takes values like 1.44, 1.50, 1.56, we might set $t_1 = s_1/4$ and $t_2 = s_2/0.04$.

Next, we order the standardized statistics by magnitude so that $t_{(1)} \leq \cdots \leq t_{(k)}$. Denote the corresponding hypotheses as $H_{(1)}, \ldots, H_{(k)}$. The probability that at least one of these statistics will be significant by chance alone at the $\alpha$ level is $1 - (1 - \alpha)^k \approx k\alpha$. But once we have rejected one hypothesis (assuming it was false), there will only be $k - 1$ true hypotheses to guard against rejecting.

Begin with i= 1 and

1. Repeatedly resample the data, (with or without replacement), estimating the cutoff value $\phi(\alpha, k - i + 1)$ such that $\alpha = \Pr\{T(k - i + 1) \leq \phi(\alpha, k - i + 1)\}$ where $T(k - i + 1)$ is the largest of the $k - i + 1$ test statistics $t_{(1)} \cdots t_{(k-i+1)}$ for a given resample.

2. If $t_{(k-i+1)} \leq \phi(\alpha, k - i + 1)$, then accept all the remaining hypotheses $H_{(1)}, \ldots, H_{(k-i+1)}$ and STOP.

Otherwise, reject $H_{(k-i+1)}$, increment $i$, and RETURN to step 1.

## 9.5   Summary

In this chapter, you learned the essentials of multivariate analysis for one-, two-, and $k$-sample comparisons and applied them to repeated measures on the same subject. You learned how to detect clustering in time and space and to validate clustering models. You used the generalized quadratic form in its several guises including Mantel's $U$ and Mielke's multiresponse permutation procedure to work through applications in epidemiology, and archeology. And you learned how to combine the results of multiple, simultaneous analyses.

## 9.6   To Learn More

Blair et al. [1994] and van-Putten [1987] review alternatives to Hotelling's $T^2$; Boyett and Shuster [1977] consider its medical applications. Extensions to other experimental designs are studied by Pesarin [1997; 2001] and Barton and David [1961].

Mantel's $U$ has been rediscovered frequently, often without proper attribution (see Whaley, 1983). Empirical power comparisons between MRPP rank tests and with other rank tests are made by Tracy and Tajuddin [1985] and Tracy and Khan [1990].

The generalized quadratic form has seen widespread application in anthropology [Williams-Blangero 1989], archaeology [Klauber 1971], ecology [Bryant 1977; Douglas and Endler 1982; Highton 1977; Levin 1977; Mueller and Altenberg 1985; Royaltey, Astrachen, and Sokal 1975; Ryman et al. 1980; Syrjala 1996], earth science [Mieleke 1991], education [Schultz and Hubert 1976], epidemiology [Alderson and Nayak 1971; Fraumeni and Li 1969; Glass and Mantel 1969; Glass et al. 1971; Klauber and Mustacchi 1970; Kryscio et al. 1973; Mantel and Bailar 1970; Merrington and Spicer 1969; Siemiatycki and McDonald 1972; Smith and Pike 1976], forestry [Cade, 1997], geography [Cliff and Ord 1971, 1981; Hubert, Golledge, and Costanzo 1982; Hubert et al. 1984], management science [Graves and Whinston 1970], meteorology [Wong, Chidambaram, and Mielke 1983], ornithology [Cade and Hoffman 1993], paleontology [Marcus 1969], psychology [Hubert and Schultz 1976], sociology [Hubert and Baker 1977, 1978], and systematics [Dietz 1983; Gabriel and Sokal 1969; Selander and Kaufman 1975; Sokal 1979]. Siemiatycki [1978] considered various refinements.

Blair, Troendle, and Beck [1996], Troendle [1995], and Westfall and Young [1993] expand on the use of permutation methods to analyze multiple hypotheses; also, see the earlier work of Shuster and Boyett [1979], Ingenbleek [1981], and Petrondas and Gabriel [1983]. Simultaneous comparisons in contingency tables are studied by Passing [1984]. For additional insight into the analysis of repeated measures see Zerbe and Murphy [1986].

## 9.7    Exercises

1. You are studying a new tranquilizer you hope will minimize the effects of stress. The peak effects of stress manifest themselves between 5 and 10 minutes after the stressful incident, depending on the individual. To be on the safe side, you've made observations at both the 5- and 10-minute marks.

   | Subject | Prestress | 5-minute | 10-minute | Treatment |
   |---------|-----------|----------|-----------|-----------|
   | A | 9.3 | 11.7 | 10.5 | Brand A |
   | B | 8.4 | 10.0 | 10.5 | Brand A |
   | C | 7.8 | 10.4 | 9.0 | Brand A |
   | D | 7.5 | 9.2 | 9.0 | New drug |
   | E | 8.9 | 9.5 | 10.2 | New drug |
   | F | 8.3 | 9.5 | 9.5 | New drug |

   How would you correct for the prestress readings? Is this a univariate or a multivariate problem? List possible univariate and multivariate test statistics. Perform the permutation tests and compare the results.

**2.** Show that Pitman's correlation is a special case of Mantel's $U$.

**3.** Sixteen seedlings were planted in two trays, eight per tray. The first tray contained a new fertilizer but was otherwise the same as the second tray. The heights of the surviving plants were compared after two weeks. Use Pesarin's test to analyze the results.

      Tray 1:    5", 4", 7", 7.5", 6", 8"
      Tray 2:    5", 4.5", 3", 6"

# CHAPTER 10

# Model Building

In this chapter, you advance from qualitative hypotheses to quantitative models linking cause and effect. As always, you'll begin with your reports, listing and, preferably, graphing your anticipated results. From these graphs you will derive formal models combining deterministic and stochastic (that is, random) elements, then use the methods developed in Chapter 4 to estimate the values of model parameters and test hypotheses about them.

## 10.1 Picturing Relationships

Picturing a relationship in physics is easy. A funeral procession travels along the freeway at a steady 55 miles an hour, so that when we plot its progress on a graph of distance traveled versus time, the points all fall along a straight line as in Figure 10.1.

A graph of my own progress when I commute to work looks a lot more like Figure 10.2; this is because I occasionally go a little heavy on the gas pedal, while at other times traffic slows me down. Put a highway patrol car in the lane beside me and a graph of my progress would look much like that of the funeral procession. The underlying pattern is the same in each case—the distance traversed increases with time. But in real life, random fluctuations in traffic result in accelerations and decelerations in the curve.

The situation is similar but much more complicated when we look at human growth as in Figure 10.3: A rapid rate of growth for the first year, slow steady progress for the next 10 to 14 years and then, almost overnight it seems, we're a different size. Inflection points in the growth curve are different for different individuals. My ex-wife was 5'6" tall at the end of grade six, and 5'7" tall when she married me. I was 4'10" at the end of grade six, 5' at the beginning of grade ten, and 5'10" when she married me. The pattern was the same for both of us, but the timing of our growth spurts differed.

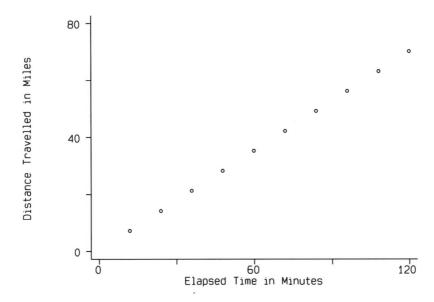

**Figure 10.1.**  Graphing the progress of a funeral procession.

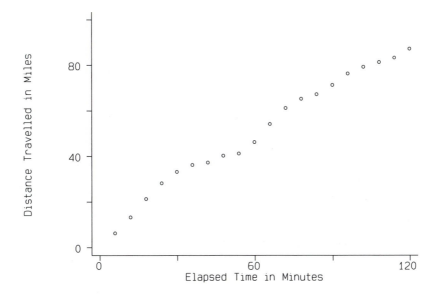

**Figure 10.2.**  Graphing my progress through freeway traffic.

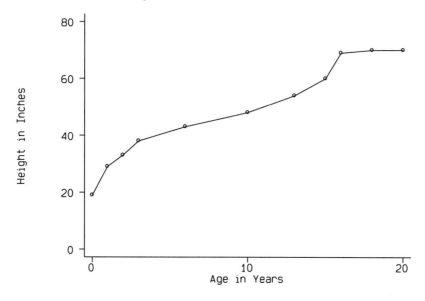

**Figure 10.3.**   Growth curve of Donny Travaglia.

These differences among individuals are why we need a way to characterize both average behavior and variation. A single number just won't do.

Before I launch an investigation, I start thinking about how and in what form I will report my results. I try to imagine the relationships I will be depicting.

Suppose, for example, I were planning to investigate the relationship between education and income. What variables should I use to quantify this relationship? years of education? annual income at age 30? Total lifetime income?

To gain further insight, I'll pretend I've done a preliminary survey of several hundred individuals. I write down guesstimates of average responses and transfer these guesstimates to a two-way plot as in Figure 10.4(a).

A positive but nonlinear relationship is depicted in Figure 10.4(a) with income rising as one completes high school and college and falling off again with each year beyond the first in graduate school. Other possibilities, such as those depicted in Figure 10.4(b), (c), and (d), include a positive linear relationship, a negative linear relationship, and no relationship between income and education.

My next step is to begin to quantify this relationship in the form of an equation. Figures 10.1, 10.2, and 10.4(b), (c), and the initial rising portion of Figure 10.4(a) all have the same underlying linear form: $Y = a + bX$, where $Y$ is the dependent variable—income or distance traveled, $X$ is the so-called "independent" variable, years of education—or time, and $a$ and $b$ are the to-be-estimated intercept and slope of the line, respectively.

Suppose for example, we were to write $I = \$20{,}000 + \$5{,}000E$ where $I$ stands for income and $E$ for years of education after high school. This relationship is depicted in Figure 10.4b. Among its implications are that without college ($E = 0$) average annual income is \$20,000, while with college completed ($E = 4$) average

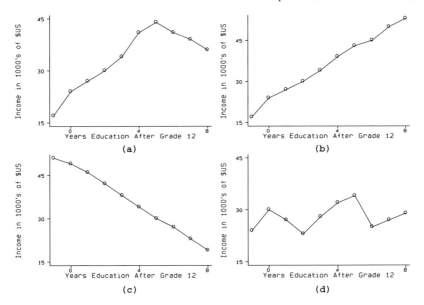

**Figure 10.4.** Various trend lines for income as a function of years of education. (a) expected, (b) rising linear, (c) falling linear, (d) no association.

income is doubled or $40,000.

## 10.2   Unpredictable Variation

If our model were perfect, then all the points and all future observations would lie exactly on a straight line. We might be able to improve the fit to the existing values by writing the dependent variable, income in the previous example, as a function of several different variables $X_1, X_2, \ldots, X_n$ each representing some characteristic that might influence future income, but, again, it's unlikely the fit would be perfect. Even with time-tested models, of the kind studied in freshman science laboratories, the observations just won't cooperate. Always, there seems to be a portion we can't explain, the result of observer error, or DNA contamination, or a hundred other factors we did not think of measuring or were unable to measure.

For simplicity, let's represent all the different explanatory variables by the single letter $X$, and again suppose that even with all these additional variables included in our model, we still aren't able to predict $Y$ exactly. A small fraction $\varepsilon$ of each observation continues to defy explanation. Our solution is to write $Y$ as a mixture of deterministic and stochastic (random) components,

$$Y = bX + \varepsilon,$$

where $X$ represents the variables we know about, $b$ is a vector of constants used to apportion the effects of the individual variables that make up $X$, and $\varepsilon$ denotes the

---

**Developing and Testing a Model**

1. Begin with the Reports. Picture Relationships.
2. Convert Your Pictures to Formal Models with both Deterministic and Stochastic Elements.
3. Plan Your Sampling Method.

   Define the Population of Interest.

   Ensure

   a) Sample is representative.
   b) Observations are independent.

---

Table 10.1. Registrants and Servings at Fawlty Towers

| Registrants | Maximum Servings | Registrants | Maximum Servings |
|---|---|---|---|
| 289 | 235 | 339 | 315 |
| 391 | 355 | 479 | 399 |
| 482 | 475 | 500 | 441 |
| 358 | 275 | 160 | 158 |
| 365 | 345 | 319 | 305 |
| 561 | 522 | 331 | 225 |

error or random fluctuation, the part of the relationship we can't quite pin down or attribute to any specific cause.

## 10.2.1   Building a Model

Imagine you are the proud owner of Fawlty Towers and have just succeeded in booking the International Order of Arcadians and Porcupine Fanciers for a weekend conference. In theory you ought to prepare to serve as many meals as the number of registrants, but checking with your fellow hotel owners, you soon discover that with no-shows and non-diners you can get by with a great many less. Searching through the records of the former owners of your hotel, you come up with the data in Table 10.1. You convert these numbers to a graph, Figure 10.5, and discover what looks almost like a linear relationship!

## 10.2.2   Estimating the Parameters

Many, many methods exist for estimating the parameters of a linear relationship. As discussed in Section 6.1.5, your choice of method will depend upon the purpose of your estimation procedure and the losses you may be subjected to should mistakes be made.

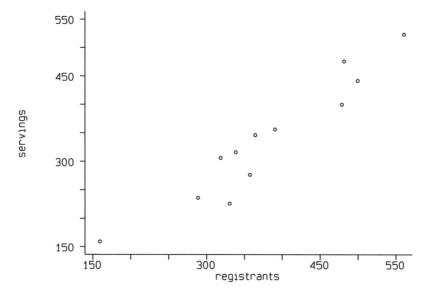

**Figure 10.5.**   Servings vs. conference registrants at Fawlty Towers.

In most instances, your purpose will be to use historical data to predict the future. This is just what the owner of Fawlty Towers is trying to do. Seeing that a relationship of the form Meals $= a + b * $ Registrants appears to exist, he wants to determine values of a and b so that when he attempts to predict the number of meals required in the future, he will minimize the losses associated with producing too few or too many meals.

The two most popular methods for determining coefficients are referred to as least-squares (LSD) and least-absolute-deviation (LAD) goodness of fit. Because they are popular, a wide selection of computer software is available to help us do the calculations.

With least-squares goodness of fit, we seek to minimize the sum

$$\sum_i (M_i - a + bR_i)^2,$$

where $M_i$ denotes the number of meals served and $R_i$ the number of registrants on the $i$th occasion. With the LAD method, we seek to minimize

$$\sum_i |M_i - a + bR_i|.$$

Those who've taken calculus, know the LSD minimum is obtained when

$$\sum_i (M_i - a + bR_i)b = 0 \quad \text{and} \quad \sum_i (M_i - a + bR_i) = 0,$$

that is, when

$$b = \frac{\text{Covariance}(RM)}{\text{Variance}(M)} = \frac{\sum (R_i - \bar{R})(M_i - \bar{M})}{\sum (M_i - \bar{M})^2}$$

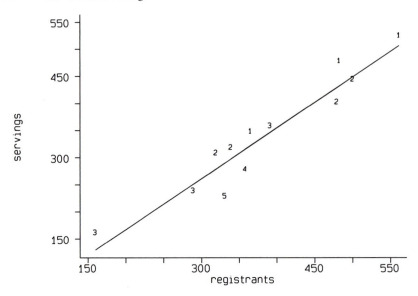

**Figure 10.6.**    Regression of servings on registrants at Fawlty Towers (Numbers refer to severity of weather conditions).

and

$$a = \bar{M} - b\bar{R}.$$

For Fawlty Towers, the LSD method yields the following equation:

$$\text{Meal Servings} = 0.94 * \text{Registrants} - 20.6.$$

Of course, there are discrepancies, also known as *residuals*, from any model, as can be seen from Figure 10.6 and as you note in Table 10.2. Is there an explanation for these discrepancies: 331 registrants, 290 meals predicted, and only 225 meals served? Well, one thing you forgot to mention to the officers of the IOAPF before you signed them up is that for most of the year the weather out Fawlty Towers way is dreadful most of the time and unpleasant the rest.[1] Perhaps, bad weather is the explanation. Digging deeper into the previous owners' records you come up with the expanded Table 10.3 in which you've assigned numerical values to the weather conditions ranging from 1-Sunny to 5-Hurricane.

Adding a term to account for the weather to your model, and again applying the methods of Chapter 4, yields the following formula

$$\text{Servings} = 0.78 * \text{Registrants} - 27 * \text{Weather} + 104.$$

The residuals as you can see from the revised table (Table 10.4) are much smaller, though, of course, the fit still is not perfect.

---

[1] The pictures you showed them of Fawlty Towers, taken on your own Hawaiian vacation, didn't help the situation. What you didn't realize and will be forced to deal with later when you come up 100 meals short is that the weather in Nova Scotia where most of those Arcadians live is awful for most of the year, also.

Table 10.2. Model I vs. Reality

| Registrants | Servings | | Residual |
| | Actual | Model I | |
|---|---|---|---|
| 289 | 235 | 251 | −15.9 |
| 391 | 355 | 347 | 8.3 |
| 482 | 475 | 432 | 42.8 |
| 358 | 275 | 316 | −40.7 |
| 365 | 345 | 322 | 22.7 |
| 561 | 522 | 506 | 15.6 |
| 339 | 311 | 298 | 17.1 |
| 479 | 399 | 429 | −30.4 |
| 500 | 441 | 449 | −8.1 |
| 160 | 158 | 130 | 28.3 |
| 319 | 305 | 279 | 25.9 |
| 331 | 225 | 290 | −65.4 |

Table 10.3. Registrants and Servings at Fawlty Towers

| Registrants | Servings | Weather | Registrants | Servings | Weather |
|---|---|---|---|---|---|
| 289 | 235 | 3 | 339 | 315 | 2 |
| 391 | 355 | 3 | 479 | 399 | 2 |
| 482 | 475 | 1 | 500 | 441 | 2 |
| 358 | 275 | 4 | 160 | 158 | 3 |
| 365 | 345 | 1 | 319 | 305 | 2 |
| 561 | 522 | 1 | 331 | 225 | 5 |

Table 10.4. Model II vs. Reality

| Registrants | Servings | | Residual |
| | Actual | Model II | |
|---|---|---|---|
| 289 | 235 | 250 | −14.5 |
| 391 | 355 | 329 | 25.6 |
| 482 | 475 | 455 | 20.2 |
| 358 | 275 | 277 | −1.6 |
| 365 | 345 | 363 | −18.1 |
| 561 | 522 | 517 | 5.3 |
| 339 | 311 | 316 | −0.7 |
| 479 | 399 | 425 | −26.4 |
| 500 | 441 | 442 | −0.9 |
| 160 | 158 | 148 | 9.6 |
| 319 | 305 | 300 | 4.9 |
| 331 | 225 | 228 | −3.4 |

## 10.2.3   Testing for Significance

To test whether the coefficients of the predictors are significantly different from zero, we simply apply the test for bivariate correlation described in Section 5.3. For $b = \rho\sigma_y/\sigma_x$; thus $b$ is zero if and only if $\rho$ is zero. Alternatively, we could permute the $x$'s while leaving the $y$'s fixed, then estimate $a$ and $b$ by either LS or LAD methods to obtain permutation distributions for the two parameters. A test based on this distribution is subject to the restriction that the $x$'s are exchangeable under the null hypothesis.

To test whether $b$ takes some specific nonzero value, $b = 100$, say, so that $y = a + 100x$, we would determine the coefficient $a$ by LAD means, then apply the one-sample tests of Section 3.6. to the residuals $r = y - a - 100x$. By doing this repeatedly for various values of $b$, we can derive confidence limits for $b$.

As shown in Section 5.4.1, we could take a similar approach using the bootstrap subject to the assumption that the residuals are independent and identically distributed. We may also obtain interval estimates by taking bootstrap samples of the multivariate vectors consisting of the predictor(s) $x$ and their "effect" $y$. Tests based on these estimates require the vectors to be exchangeable; the residuals need not be; thus, these tests are independent of the form (and possible inaccuracies) of the chosen model (Efron, 1982).

These resampling methods are independent of the functional relationship between the predictor(s) and the effect and are applicable to equations of the general form $G[y, x] = 0$ as well as to the more familiar linear form $y = a + bg[x]$. For an example of the bootstrap's application in a nonlinear regression, see Shimbukaro et al. [1984].

## 10.2.4   Comparing Two Regression Lines

A question arising often in practice is whether two regression lines based on two sets of independent observations have the same slope.[2] Suppose we can assume

$$y_{ij} = a_i + bx_{ij} + \varepsilon_{ij} \quad \text{for } i = 1, 2; \ j = 1, \ldots, n_i,$$

where the errors $\{\varepsilon_{ij}\}$ are exchangeable then

$$\bar{y}_{i.} = a_i + b\bar{x}_{i.} + \bar{\varepsilon}_{i.}$$

Define

$$y' = \frac{1}{2}(\bar{y}_1 - \bar{y}_2); \ x' = \frac{1}{2}(\bar{x}_1 - \bar{x}_2); \ \varepsilon' = \frac{1}{2}(\bar{\varepsilon}_1 - \bar{\varepsilon}_2); \ a' = \frac{1}{2}(a_1 + a_2).$$

Define $y'_{1i} = y_{1i} - y'$ for $i = 1$ to $n_1$ and $y'_{2i} = y_{2i} + y'$ for $i = 1$ to $n_2$.
Define $x'_{1i} = x_{1i} - x'$ for $i = 1$ to $n_1$ and $x'_{2i} = x_{2i} + x'$ for $i = 1$ to $n_2$.
Then $y'_{ij} = a' + bx'_{ij} + \varepsilon'_{ij}$ for $i = 1, 2; \ j = 1, \ldots, n_i$.

---

[2]This is quite different from the case of repeated measures where a series of dependent observations are made on the same subjects (Good 2000, pp. 90–93) over a period of time.

Two cases arise. If the original predictors were the same for both sets of observations, that is, if $x_{1j} = x_{2j}$ for all $j$, then the errors $\{\varepsilon'_{ij}\}$ are exchangeable and we can apply the method of matched pairs as in Section 3.7. Otherwise, we need to proceed as follows: First, estimate the two parameters $a$' and $b$ by least-squares means. Use them to derive the transformed observations $\{y'_{ij}\}$. Then test the hypothesis that $b_1 = b_2$ using a two-sample comparison as in Section 3.2. If the original errors were from a symmetric distribution and were exchangeable, then the errors $\{\varepsilon'_{ij}\}$ though not independent are exchangeable and this test is exact.

Alternately, we may know two curves are parallel, but suspect that they are not coincident. This problem is similar to that of the two-sample comparison, the difference from Section 3.1 being that we wish to increase the power of the test by correcting for the effects of various covariates. This problem also serves to illustrate some of the major differences between the permutation and the bootstrap approach.

Our solutions by resampling methods requires us to take observations from the two populations for the same set of values of the covariates $\{X_i; i = 1, \ldots, n\}$. Given that,

$$Y_{1i} = a_1 + f[X_i] + \eta_{1i} \quad \text{for } I = 1, \ldots, n$$
$$Y_{2i} = a_2 + f[X_i] + \eta_{2i} \quad \text{for } I = 1, \ldots, n,$$

where the $\{\eta_{ji}\}$ are independent random values each of whose expected value is zero.

To test the null hypothesis that $a_1 = a_2$, our statistic for the permutation test is that for matched pairs, $S = \sum(Y_{1i} - Y_{2i})$, where we perform $n$ independent permutations, one for each pair. The permutation test is exact if we can assume the $\{\eta_{ji}\}$ are identically distributed.

The statistic for the bootstrap derived by Hall and Hart [1990] is

$$S = \Big[\sum_{j=0}^{n-1}\Big(\sum_{i=j+1}^{j+m} D_i\Big)^2\Big]\Big[n\sum_{i=1}^{n-1}(D_{i+1} - D_i)^2\Big]^{-1},$$

where

$$D_i = Y_{1i} - Y_{2i} - n^{-1}\sum_{i=1}^{n}(Y_{1i} - Y_{2i}) \quad \text{for } 1 \le i \le n;$$

$$D_i = D_{i-n} \quad \text{for } n + 1 \le i \le n + m,$$

$m$ equals the integral part of $np$, with $0 \le p \le 1$.

The complexities of this statistic are occasioned by the need to Studentize to obtain asymptotically exact significance levels and the introduction of m necessary to the proof that the results are asymptotically exact. The bootstrap test requires no additional assumption beyond the original one of the independence of the errors $\{\eta_{ji}\}$.

## 10.3   Limitations of Regression

Studies of a broad range of phenomena lead to the following general conclusions:

Every relationship has both a linear and a nonlinear portion with the non-linear portion becoming evident for both extremely large and extremely small values. While a regression equation may be used for interpolation within the range of known values, we are on shaky ground if we try to extrapolate, to make predictions for conditions not previously investigated. For example, if we try to apply the equation we derived for Fawlty Towers when Weather $= 1$ and Registrants $= 100$, we would estimate a need for 150 meals. One hundred fifty meals for 100 guests! I don't think so.

Relationships are not unique. If a relationship exists between two variables $X$ and $Y$, then a relationship also exists between $Y$ and any monotone function of $X$. If we can fit the curve I: $Y = A + BX + \varepsilon$, we also can fit the relationships II: $Y = A + B \log[X] + \varepsilon$, and III: $Y = A + BX + CX^2 + \varepsilon$. It can be very difficult to determine which model if any is the "correct" one. At least two contradictory rules apply:

- the more parameters the better the fit; thus, in this example, Model III is to be preferred to Model I.

- the simpler, more straightforward model is more likely to be correct; thus, in this example, Model I is to be preferred to Model III.

Correlations can be deceptive. Variable $X$ can appear to be the cause of variable $Y$, can have a statistically significant correlation with $Y$, solely because both $X$ and $Y$ are dependent on a third variable $Z$.

The goodness-of-fit criterion we use to estimate the values of our model parameters says the best model is the one that minimizes the sum of squares for the historical data $\sum(Y_{\text{observed}} - Y_{\text{model}})^2$. But minimizing this sum of squares is no guarantee that when we continue to gather data, we will minimize the sum $\sum(Y_{\text{observed}} - Y_{\text{predicted}})^2$ based on the square of the difference between what we observe in the future and what our model predicts. If you are a businessman whose objective is to predict market response, this distinction can be critical.

### 10.3.1   Local Regression

We can overcome several of these objections by using a parameter-free technique called local regression. In its most primitive form, given a value of the predictor $x = x_i$, we predict $y$ will be the mean of all the values of $y$ we observed previously when $x = x_i$. If we have several hundred thousand observations, this method might prove satisfactory, but with small data sets such as that for Fawlty Towers, the resultant curve, while a near-perfect fit to the observed data, jumps about so as to suggest it would be unreliable for prediction purposes. We can increase its reliability by smoothing it and flattening out the jumps. One such smoothing method uses nearest neighbors: Suppose we order the $\{x_i\}$ from smallest to largest so that $x_{(1)} \leq x_{(2)} \leq \cdots \leq x_{(n)}$. Define $y_{(j)}$ to be the mean of the $y$ values observed when $x = x_{(j)}$. Our nearest-neighbor predictor when $x = x_{(j)}$ is $y = (y_{(j-1)} + 2y_{(j)} + y_{(j+1)})/4$. See Pagan and Ullah [1999; p. 81ff] for a description of other, more complex smoothing methods.

## 10.4   Validation

How can we be confident of our ability to predict the future when all we have to work with is the past? We have no choice but to make use of the same data to develop the model and to validate it. We need either develop a *metric* and determine its permutation distribution or apply a third resampling technique known as *cross-validation*. In what follows we consider each of these approaches in turn.

### 10.4.1   Metrics

With the availability of more and more powerful computers, scientists are addressing problems involving hundreds of variables such as weather prediction and economic modeling. In many instances, the computer has taken the place of a laboratory. Where else could you emulate a thousand years of stellar evolution or study over and over the effects of turbulence on an aircraft landing?

Whether the data are real or simulated, we need to analyze our results. To distinguish between close and distant, between a good fitting model and a poor one, we first need a metric, and then a range of typical values for that metric. A metric $m$ defined on two points $x$ and $y$, has the following properties:

$$m(x, y) \geq 0,$$
$$m(x, x) = 0,$$
$$m(x, y) \leq m(x, z) + m(z, y),$$

which corresponds to most of the common ways in which we measure distance. The third property of a metric, for example, simply restates that the shortest distance between two points is a straight line.

A good example is the standard Euclidian metric used to measure the distance between two points $x$ and $y$ whose coordinates in three dimensions are $(x_1, x_2, x_3)$ and $(y_1, y_2, y_3)$

$$\sqrt{(x_1 - y_1)^2 + (x_2 - y_2)^2 + (x_3 - y_3)^2}.$$

This metric can be applied even when $x_i$ is not a coordinate in space but the value of some variable like blood pressure or laminar flow or return on equity. Hotelling's $T^2$, defined in Chapter 9, is a example of such a metric.

As in Chapter 9, the next step is to establish the range of possible values that such a metric might take by chance, using either the bootstrap or permutations of the combined set of theoretical and observed values. If the value of this metric for the original observations is an extreme one, we should reject our model and begin again.

## Nearest Neighbors

Suppose we have a set of $N$ observations and a second set of $N$ simulated values. For each point in the set of theoretical values, Set 1, record the number of points among its three nearest neighbors (nearest in terms of the Euclidian metric) which are also in Set 1; sum the number of points so recorded. Call this sum $S_0$.

To determine whether $S_0$ is an extreme value, compute $S$ for rearrangements of the combined group of theoretical and observed values. In all but the original and one other arrangement, Set 1 will consist of a mixture of observed and theoretical values. If these values are significantly different from one another, fewer nearest neighbors will be in the same set and $S$ will be smaller than $S_0$.

## Structured Exploratory Data Analysis

Karlin and Williams [1984] use permutation methods in a structured exploratory data analysis (SEDA) of familial traits. A SEDA has four principal steps:

1) The data are examined for heterogeneity, discreteness, outliers, and so forth, after which they may be adjusted for covariates (as in Chapter 9) and the appropriate transform applied (as in Appendix 1).

2) A collection of summary SEDA statistics are formed from ratios of functionals as in the example that follows.

3) The SEDA statistics are computed for the original family trait values and for reconstructed family sets formed by permuting the trait values within or across families.

4) The values of the SEDA statistics for the original data are compared with the resulting permutation distributions.

As one example of a SEDA statistic, consider the OBP, the Offspring-Between-Parent SEDA statistic:

$$\frac{\sum_i^N \sum_j^{J_i} |O_{ij} - (M_i + F_i)/2|}{\sum_i^N |M_i - F_i|}.$$

In family $i = 1, \ldots, I$, $F_i$ and $M_i$ are trait values of the father and mother (cholesterol levels in their blood, for example), while $O_{ij}$ is the corresponding trait value of the $j$th child of those same parents, $j = 1, \ldots, J_i$.

To evaluate the permutation distribution of the OBP, consider all permutations in which the children are kept together in their respective family units, while we either

a) randomly assign to them a father and (separately) a mother, or

b) randomly assign to them an existing pair of spouses.

The second of these methods preserves the spousal interaction. Which method a geneticist might choose will depend upon the alternative(s) of interest.

It would be difficult if not impossible to derive the distribution of this statistic by mathematical analysis. To obtain the permutation distribution for the OBP statistic, we merely need to substitute its formula for the statistic in our sample programs in Appendix 1.

### Goodness of Fit

A metric can always be derived. Before we begin to develop our model, divide the data at random into two parts, one of which will be used for model development and estimation, the other for validation. Our goodness-of-fit metric is

$$G = \frac{\sum_{k \in \{\text{validation}\}} (Y_{\text{observed}} - Y_{\text{predicted}})^2}{\sum_{k \in \{\text{estimation}\}} (Y_{\text{observed}} - Y_{\text{predicted}})^2},$$

where the summation in the numerator is taken over all the observations in the validation data set and the summation in the denominator is taken over all the observations in the estimation data set.

For two reasons, this ratio will almost always be larger than unity:

1. The estimation data set, not the validation set, was used to choose the variables that went into the model and to decide whether to use the original values of the variables or to employ some sort of transformation.

2. The estimation data set, not the validation set, was used to estimate the values of the model coefficients.

Divide the original data set into two parts at random a second time, but use the estimation set only to calculate the values of the coefficients. Use the same model you used before, that is, if $\log[X]$ was used in the original model, use $\log[X]$ in this new one. Compute $G$ a second time.

Repeat this resampling process several hundred times. If the original model is appropriate for prediction purposes, it will provide a relatively good fit to most of the data sets; if not, the goodness-of-fit statistic for our original estimation set will be among the largest of the values, since its denominator will be among the smallest.

A worked-through example is provided in Section 10.6.1.

## 10.4.2  Cross-Validation

Four techniques of cross-validation are in general use:

*K-fold*, in which we subdivide the data into $K$ roughly equal-sized parts, then repeat the modeling process $K$ times, leaving one section out each time for validation purposes,

*Leave-one-out*, an extreme example of $K$-fold, in which we subdivide into as many parts as there are observations. We leave one observation out of our classification procedure, and use the remaining $n-1$ observations as a training set. Repeating this procedure n times, omitting a different observation each time, we arrive at a figure for the number and percentage of observations classified correctly.

A method that requires this much computation would have been unthinkable before the advent of inexpensive readily available high-speed computers. Today, at worst, we need step out for a cup of coffee while our desktop completes its efforts.

*Jackknife*, an obvious generalization of the leave-one-out approach, where the number left out can range from one observation to half the sample.

*Delete–d*, where we set aside a random percentage $d$ of the observations for validation purposes, use the remaining $100 - d\%$ as a training set, then average over 100 to 200 such independent random samples. The goodness-of-fit statistic is an example of delete–50.

The *bootstrap* is a fifth, recommended alternative [Efron 1983].

We may use any of these methods to determine which of several competing models to adopt. For the data from Fawlty Towers, let us use the delete–50% cross-validation method, dividing our sample into two equal parts at random. We'll use the first part of the data to estimate the model parameters and the second part to check the accuracy of the resultant model or models.

For example, we might use the six pairs of observations (289,235), (391,355), (358,275), (339,315), (160,159), (319,305) to estimate the regression coefficients for each of the competing models, and the deviation when we try to fit the remaining six pairs to determine the prediction errors.

We repeat this process 20 or 30 times—selecting six points, estimating the regression coefficients, computing the prediction errors. Our model of choice is the one that minimizes the average prediction error.

## 10.5   Prediction Error

For most hotel owners, John Fawlty excepted, a single figure would not be enough; a prediction interval is needed. Averages are often exceeded and the sight of six or seven conference participants fighting over a single remaining pork cutlet can be a bit unsettling. In this section, we bootstrap not once, but twice to obtain the desired result.

Consider the more general regression problem where given a vector $X$ of observations, we try to predict $Y$. Suppose we've already collected a set of independent identically distributed observations $w = \{w_i = (x_i, y_i), i = 1, \ldots, n\}$ taken from the multidimensional distribution function $F$. The data in Table 10.1, with $X$ the number of registrants and $Y$ the number of meals, provides one example. Given a new value for the number of registrants $x_{n+1}$ we use our regression equation based on $w$ to derive an estimate $m[x_{n+1}, w]$ of the number of meals we can expect to serve. Any difference between the estimate and the actual number of meals served could result in a loss. This loss could be something as simple as the cost of preparing an unnecessary meal or it could be far more complicated including with the cost of a replacement meal obtained from a nearby hotel, and the costs of lawsuits brought by thoroughly irritated porcupine fanciers.

The prediction error $L(m_{n+1}, m[x_{n+1}, w])$ should be averaged over all possible outcomes $(x_{n+1}, y_{n+1})$ we might draw from $F$ in the future. Call this error

$E[w, F]$. We don't know what $F$ is, but we do have an estimate, $F'$ based on the data $w$ we've collected so far. The *apparent error rate* is the average of the losses over all the pairs of values $(x_i, y_i)$ we have observed:

$$E[w, F'] = \sum_i L(m_i, m[x_i, w])/n.$$

To estimate the true error, we treat the observations as if they were the population, and a series of bootstrap samples as if they were the observations. Each bootstrap sample consists of pairs $\{w_i^*\}$ drawn without replacement from $w$. Our plug-in estimate of error based on a bootstrap sample is

$$E[w^*, F'] = \sum_i L(m_i, m[x_i, w^*])/n.$$

To obtain this estimate, we must compute the regression coefficients a second time using $w^*$ in place of $w$ in our calculations, although we continue to average the errors over the original observations. As results based on a single bootstrap sample can be quite misleading, we have to repeat the process of sampling, regression, and error determination several hundred times, then take the average of our estimates. I'm not complaining for, like you, I've got a computer to do the work. Our final estimate of prediction error is the average of the plug-in estimates over the $B$ bootstrap samples,

$$E[w^*, F^b] = \sum^B \sum_i L(m_i, m[x_i, w^*])/nB.$$

We use the distribution of errors from these same calculations to obtain an interval estimate of the prediction error.

---

**Estimating Prediction Error I**

Determine the loss function $L$.
Repeat the following several hundred times:
  Draw a bootstrap sample $w^*$ from the pairs of observation $w = \{(x_i, m_i)\}$.
  Compute the regression coefficients based on this sample $w^*$.
  Using these coefficients, compute the losses $L(m_i, m[x_i, w^*])$ for each pair in the original sample.
  Compute the mean of these losses.
Compute the mean of these means averaged over all the bootstrap samples.

---

## 10.5.1   Correcting for Bias

When we estimate a population mean using the sample mean, the result is unbiased, that is, the mean of the means of all possible samples taken from a population is the population mean. In the majority of cases, including the preceding example, estimates of error based on bootstrap samples are biased and tend to underestimate the actual error. If we can estimate this bias by bootstrapping a second time, we can improve on our original estimate.

This bias, which Efron and Tibshirani [1993] call the *optimism*, is the difference $E[w, F] - E[w, F^b]$ which we estimate as before by using the observations in place of the population and the bootstrap samples in place of the observations,

$$E[w^*, F^b] - E[w, F^{b*}] = \sum^B \sum_i \{L(m_i, m[x_i, w^*]) - L(m_i^*, m[x_i^*, w^*])\}/nB.$$

$L(m_i, m[x_i, w^*])$ uses the observations from the original sample and the coefficients derived from the bootstrap; $L(m_i^*, m[x_i^*, w^*])$ uses the bootstrap observations and the coefficients derived from the bootstrap. Our corrected estimate of error is the apparent error rate plus the optimism or

$$E[w, F^b] + E[w*, F^b] - E[w, F^{b*}].$$

---

**Estimating the Prediction Error II**

Determine the loss function $L$.
Solve for the regression coefficients using the original observations.
Compute apparent error rate:

$$E[w, F'] = \sum_i L(m_i, m[x_i, w])/n.$$

Repeat the following 160 times:
   Choose a bootstrap sample $w^*$ with replacement.
   Solve for regression coefficients using the bootstrap sample.
   Use these coefficients to determine apparent error for original observations

$$E[w^*, F^{b*}]$$

and the optimism
$$E[w^*, F^b] - E[w, F^{b*}].$$

Compute the average optimism.
Compute the corrected error rate.

---

## 10.6   Multivariate Relationships

In most practical situations, we need to consider and measure not merely two or three but dozens, or even hundreds, of variables simultaneously. How do we

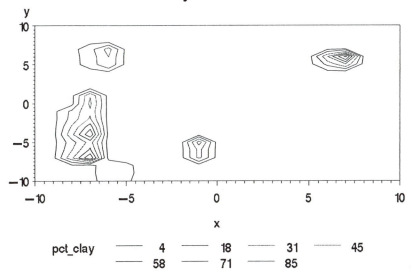

**Figure 10.7.** Clay content as a function of location. (Derived with the help of SAS/Graph™)

depict the results? One way to see three variables on a single two-way plot is to use the third variable as a plotting symbol as in Figure 1.2. Another is to utilize contours that delineate equally spaced values of the third variable; an example is Figure 10.7, prepared with the aid of SAS®. Still another is to use pseudo-perspective as in Figure 10.8.

A fourth excellent solution is the scatter-plot matrix. In Figure 10.9, we've used the Stata graph-matrix routine to plot all possible two-way scatterplots of four variables—median family income, housing units, median gross rent, and the population density. Each point corresponds to 1 of the 50 states.

## 10.6.1 Limitations of Multiple Regression

When we attempt to develop a model based on the values of a large number of interdependent variables, we run into additional difficulties over and above those we discussed in Section 10.4. If these variables are correlated as they will be in most cases, then the values of the coefficients as determined by least-squares or LAD methods and the relative import of each variable will depend upon the order in which the variables are introduced into the model. As a result the predictive value of the model

$$Y = a_0 + a_1 X_1 + a_2 X_2 + a_i X_i + \ldots + a_n X_n$$

may differ markedly from that of the model

$$Y = a'0 + a'_i X_i + a'_1 X_1 + a'_2 X_2 + \ldots + a'_n X_n$$

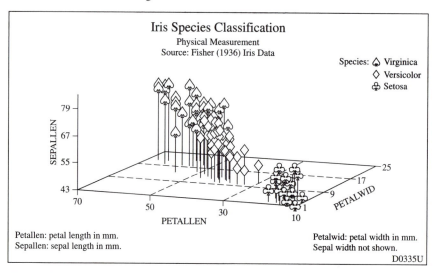

**Figure 10.8.** Representing three variables in two dimensions. Iris species representation derived with the help of SAS/Graph®.

because changing the order in which we solve for the $\{a_i\}$ coefficients changes their values.[3]

When too many variables are involved, many correlations will be entirely spurious, present by chance in the sample, but not in the larger population. David Freedman [1983] generated a random sample in which each multivariate observation was composed of 50 completely independent variables. Nonetheless, when he designated one of the variables at random as the "effect," he found that 15 of the remaining variables appeared to be significantly correlated with it, at least in the statistical sense. And when he built a model for the "effect" based on those 15 "causes," the results were highly significant at the 1% level.

A further problem arises when we need to decide which *interaction* terms to include in the model. An example of an interaction would be a term of the form $aX_1X_2$. Interaction correct for shifts in the relationships among the variables as the values of the variables are changed. An excellent example is the effect of age on the relationship between a person's height and weight. During the growth period between 1 and 14 years, height and weight are closely correlated. But over the age of 20, the only body dimension that increases with weight is width. A possible model relating height $H$ to weight $W$ would be $H = a_0 + a_1 W A$, where $A$ is a variable that takes various values depending on age.

If we have $K$ independent variables, we will have $K(K-1)$ interaction terms of the form $X_iX_k$, $K(K-1)(K-2)$ of the form $X_iX_kX_m$, and so forth. If we're not careful, we'll end up with more terms than we have observations with which to estimate them. Regression methods alone cannot make the determination.

---

[3]This situation is not helped by the unfortunate terminology used in the literature in which $Y$ is called the "dependent variable" and $X_1$, $X_2$, etc., the "independent." They are all *interdependent*.

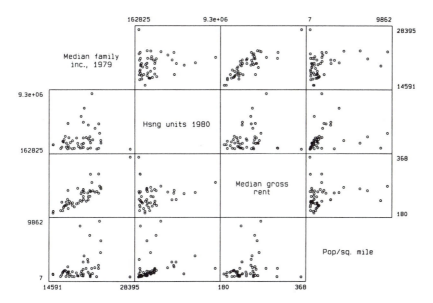

**Figure 10.9.**   Matrix of scatter plots. Prepared with Stata®.

---

### Guidelines for Model Building

Your objective in modeling is prediction, not goodness of fit:

- Use the minimum number of predictors—over-fitted models are numerically unstable.

- Use automated methods only to produce several apparently good models that can be investigated further.

- Include only predictors with which a plausible cause and effect relationship can be established.

- Replace multiple highly correlated (collinear) predictors with an average or some other linear combination.  This procedure recommends itself when a coefficient that should be positive (negative) has the opposite sign.

- When investigating models for several similar products, select the model that fits all the products reasonably well, rather than trying to find the best model for each product.  (Any exception should be justified on a cause and effect basis.)

## 10.6.2  Applied Marketing Research

We illustrate many of the points raised in the preceding sections with the following example provided through the courtesy of Megaputer Intelligence, Inc. and distributed freely along with the demonstration version of their data mining and regression software, PolyAnalyst™.

In order to optimize an advertising campaign by directing it toward the best potential customers, a study of consumers' attitudes, interests, and opinions was commissioned. The questionnaire consisted of 31 statements covering a variety of dimensions, including consumers' attitudes towards risk, foreign-made products, product styling, spending habits, self-image, and family. The final question concerned purchasing the product itself.

Respondents used a nine-point Likert scale, where a value of "1" meant they definitely disagreed with a statement, and "9" meant they definitely agreed. A total of 400 respondents were obtained from the mailing lists of *Car and Driver, Business Week*, and *Inc.* magazines, who were then interviewed at their homes by an independent surveying company.

The survey consisted of the following 31 statements:

1. I am in very good physical condition.

2. When I must choose, I dress for fashion, not comfort.

3. I have more stylish clothes than most of my friends.

4. I want to look a little different from others.

5. Life is too short not to take some gambles.

6. I am not concerned about the ozone layer.

7. I think the government is doing too much to control pollution.

8. Basically, society today is fine.

9. I don't have time to volunteer for charities.

10. Our family is not too heavily in debt today.

11. I like to pay cash for everything I buy.

12. I pretty much spend for today and let tomorrow bring what it will.

13. I use credit cards because I can pay the bill off slowly.

14. I seldom use coupons when I shop.

15. Interest rates are low enough to allow me to buy what I want.

16. I have more self-confidence than most of my friends.

17. I like to be considered a leader.

18. Others often ask me to help them out of a jam.

19. Children are the most important thing in a marriage.

20. I would rather spend a quiet evening at home than go out to a party.

21. Foreign-made cars can't compare with American-made cars.

22. The government should restrict imports of products from Japan.

23. Americans should always try to buy American products.

24. I would like to take a trip around the world.

25. I wish to leave my present life and do something entirely different.

26. I am usually among the first to try new products.

27. I like to work hard and play hard.

28. Skeptical predictions are usually wrong.

29. I can do anything I set my mind to.

30. Five years from now, my income will be a lot higher than it is now.

31. I would consider buying the product.

The original file consisted of 400 records, one for each respondent. The first field in each record is the $31^{st}$ question and is labeled Attitude. The rest of the fields are abbreviated as $q1$, $q2$, and so on. Using the twofold method of cross-validation, we divide this file at random into two files each containing 200 records. We use the first of these files to develop a model and to estimate the parameters and the second to validate and discriminate among the various competing models.

In our first attempt at model development, we utilize all the statements as predictors of attitude with the results shown in Table 10.5. Two of the statements $q5$ and $q24$ stand out in Column 5 of the table as significant predictors, while five others $q7$, $q8$, $q9$, $q17$, and $q26$ have small enough values for $p > |t|$ to suggest they might also be of importance.[4]

How good is this model? Two measures of goodness of fit are in common use, the first measures the proportion of sample variation that can be explained by the model, termed $R^2$. The second measure adjusts for the number of predictors via the formula $1 - (1 - R^2)(n - 1)/(n - k)$, where $n$ is the number of observations and $k$ is the number of right-hand-side (predictor) variables including the constant term or *intercept*. In this case $R^2$ is 0.5508, but the adjusted $R^2$ is only 0.4710.

Mindful that these results may have depended on the order in which the computations were made, $q1$, then $q2$, and so forth, we proceed instead in stepwise fashion, first looking at all the models of the form attitude $= a + bq_i$, then adding a variable at a time to obtain the best two-variable model, the best three-variable model and so on until the additional terms are not significant.[5] The result of this forward stepwise approach is shown in Table 10.6. $R^2$ is 0.5053, less than the all-variables model, but the adjusted $R^2$ is 0.4872.

This model takes the form (I)

$$\text{Attitude} = -3.97 + 0.56q5 + 0.65q24 + 0.39q2 + 0.33q29$$

---

[4]These probabilities are based on the often-far-from-true assumption of normally distributed residuals. Unfortunately, no generally accepted resampling method for determining the statistical significance of individual model coefficients exists at present (see Good, 2000; pp. 127–130, for details).

[5]I elected to use a significance level of 10% in the present case.

Table 10.5. Result of Regressing Attitude on all Variables Simultaneously[6]

|      | Coef | Std. Err | t | $p > \lvert t \rvert$ | [95% Confidence Interval] | |
|------|------|----------|---|---------|---------------------------|---|
| q1   | .068703   | .2261276  | 0.304  | 0.762 | −.3776955 | .5151015 |
| q2   | .2987579  | .2306615  | 1.295  | 0.197 | −.1565912 | .7541069 |
| q3   | −.0180748 | .1180014  | −0.153 | 0.878 | −.2510215 | .2148718 |
| q4   | .1639507  | .1290876  | 1.270  | 0.206 | −.0908812 | .4187827 |
| q5   | .5383143  | .0934047  | 5.763  | 0.000 | .353924   | .7227047 |
| q6   | −.0672141 | .1521273  | −0.442 | 0.659 | −.3675287 | .2331004 |
| q7   | .2784833  | .1725207  | 1.614  | 0.108 | −.0620898 | .6190565 |
| q8   | −.4563814 | .2700595  | −1.690 | 0.093 | −.989506  | .0767431 |
| q9   | .3715754  | .2428929  | 1.530  | 0.128 | −.1079197 | .8510704 |
| q10  | .3236199  | .2423974  | 1.335  | 0.184 | −.1548969 | .8021367 |
| q11  | −.3136613 | .2542119  | −1.234 | 0.219 | −.8155011 | .1881785 |
| q12  | .161744   | .2467316  | 0.656  | 0.513 | −.3253289 | .648817  |
| q13  | .0159514  | .2307885  | 0.069  | 0.945 | −.4396483 | .4715512 |
| q14  | .1103155  | .2198704  | 0.502  | 0.617 | −.3237308 | .5443618 |
| q15  | −.1547419 | .1566642  | −0.988 | 0.325 | −.4640126 | .1545289 |
| q16  | .2663312  | .1613665  | 1.650  | 0.101 | −.0522224 | .5848849 |
| q17  | −.3611526 | .1879341  | −1.922 | 0.056 | −.7321534 | .0098482 |
| q18  | .1546187  | .135754   | 1.139  | 0.256 | −.1133733 | .4226108 |
| q19  | .1407303  | .1138782  | 1.236  | 0.218 | −.0840768 | .3655374 |
| q20  | .0091916  | .1179664  | 0.078  | 0.938 | −.223686  | .2420691 |
| q21  | .1696044  | .2390439  | 0.710  | 0.479 | −.3022923 | .6415011 |
| q22  | .0540831  | .225144   | 0.240  | 0.810 | −.3903738 | .4985401 |
| q23  | −.2075308 | .177395   | −1.170 | 0.244 | −.5577264 | .1426648 |
| q24  | .7365906  | .2111484  | 3.488  | 0.001 | .3197624  | 1.153419 |
| q25  | −.1308863 | .2145273  | −0.610 | 0.543 | −.5543846 | .2926121 |
| q26  | −.1831707 | .1177634  | −1.555 | 0.122 | −.4156474 | .049306  |
| q27  | .0561623  | .0877079  | 0.640  | 0.523 | −.1169819 | .2293065 |
| q28  | −.1633266 | .3199292  | −0.511 | 0.610 | −.794899  | .4682457 |
| q29  | .3767823  | .3153598  | 1.195  | 0.234 | −.2457696 | .9993343 |
| q30  | .1500474  | .1506405  | 0.996  | 0.321 | −.147332  | .4474269 |
| cons | −5.44761  | 1.418234  | −3.841 | 0.000 | −8.247346 | −2.647874 |

Table 10.6. Result of Forward Stepwise Regression ($p = 0.10$)

|      | Coef. | Std. Err. | t | $p > \lvert t \rvert$ | [95% Confidence Interval] | |
|------|-------|-----------|---|---------|---------------------------|---|
| q5   | .5636339  | .0883218  | 6.382  | 0.000 | .3894282  | .7378396  |
| q24  | .6466732  | .1086146  | 5.954  | 0.000 | .4324422  | .8609043  |
| q2   | .3907244  | .0848398  | 4.605  | 0.000 | .2233867  | .5580622  |
| q29  | .3323193  | .0854961  | 3.887  | 0.000 | .163687   | .5009516  |
| q7   | .1681985  | .0815285  | 2.063  | 0.040 | .007392   | .3290049  |
| q26  | −.2473945 | .1052341  | −2.351 | 0.020 | −.4549579 | −.0398312 |
| q12  | .1712023  | .0863257  | 1.983  | 0.049 | .0009338  | .3414708  |
| cons | −3.976332 | .8143577  | −4.883 | 0.000 | −5.582568 | −2.370096 |

Table 10.7. Result of Regression with Variables Selected by PolyAnalyst

|  | Coef. | Std. Err | $t$ | $p > |t|$ | [95% Confidence Interval] | |
|---|---|---|---|---|---|---|
| q5 | .5800092 | .0880265 | 6.589 | 0.000 | .4063918 | .7536266 |
| q24 | .6065941 | .107878 | 5.623 | 0.000 | .393823 | .8193652 |
| q2*q2 | .044684 | .0095065 | 4.700 | 0.000 | .0259341 | .063434 |
| q29*q29 | .0377103 | .0093062 | 4.052 | 0.000 | .0193553 | .0560653 |
| q7 | .1846818 | .0820207 | 2.252 | 0.025 | .0229098 | .3464537 |
| q26 | −.2072796 | .103602 | −2.001 | 0.047 | −.411617 | −.0029421 |
| cons | −1.975125 | .6721693 | −2.938 | 0.004 | −3.300866 | −.6493844 |

$$+ 0.17q7 - 0.25q26 + 0.17q12.$$

Using a proprietary data mining method, PolyAnalyst suggested a better fit would be obtained using $q5$, $q24$, $q2 * q2$, $q29 * q29$, $q7$, and $q26$ as predictors. The results are tabulated in Table 10.7. $R^2$ is 0.4988 and the adjusted $R^2$ is 0.4833. This model takes the form (II)

$$\text{Attitude} = -1.97 + 0.58q5 + 0.61q24 + 0.45q2 * q2$$
$$+ 0.38q29 * q29 + 0.18q7 - 0.21q26.$$

Which is the best model? If goodness of fit were the criterion, then the all-variables model would be the runaway winner. But we wish to predict the behavior of consumers in the population at large, consumers who are not part of the present study. Fortunately, we have a second, validation sample at hand. We will fit each of our models to this sample and see which one yields the smallest value of the sum of the squared residuals, $S = \sum_{i=1}^{200}(A_i - P_i)^2$, where $A_i$ is the observed value of the attitude and $P_i$ is the predicted.

For the validation data set, our first model (I) yields a value of $S =2.83$. Our second model (II) yields a smaller value $S =2.68$, and it is this latter model we should use.

What if we had begun with the validation set to obtain our estimates and again used a forward stepwise regression? The results are shown in Table 10.8. We see our old friends statements $q5$, $q24$, $q29$, and $q2$ but in place of $q7$, $q26$, and $q12$ which went into Model I we find $q17$, $q25$, and $q15$. $R^2$ is 0.6805 and even the adjusted $R^2$ is 0.6689, both deceptively large values considering that the model that was appropriate for one of our randomly selected data sets is not appropriate for the other. A possible explanation for the discrepancy is that the population from which the validation and estimation data sets are drawn consists of several distinct groups of buyers with differing relationships among

---

[6]"Coef" is the estimated value of the coefficient for that term in the model, "Std. Err" is the standard error of that estimate, "$t$" the magnitude of the $t$-statistic used to assess the statistical significance of the coefficient, "$p > |t|$" its significance level, and the remaining columns hold the upper and lower confidence bounds for the coefficient. These bounds as well as the significance level are based on a normal approximation that is valid only to the extent that the error terms are independent identically normally distributed.

Table 10.8. Result of Regression with Validation Data Set

|  | Coef. | Std. Err | t | p > \|t\| | [95% Confidence Interval] | |
|---|---|---|---|---|---|---|
| q5 | .6333929 | .0732555 | 8.646 | 0.000 | .4889041 | .7778817 |
| q24 | .4032924 | .1665758 | 2.421 | 0.016 | .0747389 | .7318459 |
| q29 | .3614619 | .0681858 | 5.301 | 0.000 | .2269725 | .4959513 |
| q2 | .3176424 | .0751629 | 4.226 | 0.000 | .1693914 | .4658934 |
| q17 | .1645959 | .074077 | 2.222 | 0.027 | .0184866 | .3107052 |
| q25 | .3031125 | .1562058 | 1.940 | 0.054 | −.0049872 | .6112121 |
| q15 | .1385831 | .0762847 | 1.817 | 0.071 | −.0118806 | .2890467 |
| cons | −5.56649 | .7242478 | −7.686 | 0.000 | −6.994997 | −4.13799 |

the variables within each group. Differences in the relative proportions of the groups in the two samples would explain why a single model will not serve for both. Several distinct models, at least one per segment, may be required. Methods for identifying and distinguishing the different population segments are the topic of Section 10.7.

Table 10.9. Comparison of Regression Models

| Model | Obs | S | Std. Dev. | Min Dev. | Max Dev. |
|---|---|---|---|---|---|
| I | 200 | 2.83 | 3.69 | .00002 | 18.53 |
| II | 200 | 2.69 | 3.59 | .00006 | 21.17 |
| GofF | 200 | 2.31 | 3.217 | .00045 | 18.22 |

## 10.6.3    Transformations

Not all relationships are linear, some are logarithmic or periodic or based on a square or square root. You may need to transform the original observations before fitting a model. The primary choices lie among logarithms, polynomials, and periodic functions.

If the equation $y = ax + b$ provides a reasonable fit to the data, so too will the equation $\log[y] = \log[a] + \log[x + c]$. We may know a priori which form is preferable. If not, we need to study the residuals from the first equation. If the variance of the residuals is more or less independent of the value of the predictor $x$, then we would probably use the original non-transformed variables. But if larger values of $x$ are associated with correspondingly larger errors, then we should use the logarithmic transform before trying to fit the data. If the logarithmic transformation results in an over-correction, then we should try the square root.

Polynomials are of the form $P[x] = bx + cx^2 + dx^3$. They should be kept as simple as possible. The polynomial $dx^3$ is to be preferred to $bx + cx^2 + dx^3$ as it entails only a single unknown parameter rather than three unknowns. The units in which the highest-order term is expressed should bear a cause and effect relationship with the units of the item to be predicted.

Periodic functions such as $\sin[2\pi t/7]$ should be used as transformations when an underlying periodicity is suspected. For example, the function $\sin[2\pi t/7]$ would be used to correct for a weekly cycle in daily recordings.

## 10.7  Data Mining

*Data mining* can and should be used to select relevant variables and combinations thereof. One such technique is the computer-aided binary regression tree or CART developed by Breiman et al. [1984]. CART subdivides a set of observations into $k$ distinct classes based on the values of M categorical or ordinal variables. The idea is to pose a series of questions to a training set whose observations are already categorized. Each question subdivides the training set (or a subset thereof) into two groups of more homogeneous composition. These questions may be simple:

"Is the $k$th variable large?"

or complex:

"Is the sum of the $j$th and $k$th variables large while the $m$th variable is small?"

The results can be used in two ways: 1) to classify observations, 2) to select model variables. We illustrate both applications in what follows.

Table 10.10: Iris Data

| Spec | Sepal Length | Sepal Width | Petal Length | Petal Width | Spec | Sepal Length | Sepal Width | Petal Length | Petal Width |
|---|---|---|---|---|---|---|---|---|---|
| 1 | 5.1 | 3.5 | 1.4 | 0.2 | 2 | 6.6 | 3 | 4.4 | 1.4 |
| 1 | 4.9 | 3 | 1.4 | 0.2 | 2 | 6.8 | 2.8 | 4.8 | 1.4 |
| 1 | 4.7 | 3.2 | 1.3 | 0.2 | 2 | 6.7 | 3 | 5 | 1.7 |
| 1 | 4.6 | 3.1 | 1.5 | 0.2 | 2 | 6 | 2.9 | 4.5 | 1.5 |
| 1 | 5 | 3.6 | 1.4 | 0.2 | 2 | 5.7 | 2.6 | 3.5 | 1 |
| 1 | 5.4 | 3.9 | 1.7 | 0.4 | 2 | 5.5 | 2.4 | 3.8 | 1.1 |
| 1 | 4.6 | 3.4 | 1.4 | 0.3 | 2 | 5.5 | 2.4 | 3.7 | 1 |
| 1 | 5 | 3.4 | 1.5 | 0.2 | 2 | 5.8 | 2.7 | 3.9 | 1.2 |
| 1 | 4.4 | 2.9 | 1.4 | 0.2 | 2 | 6 | 2.7 | 5.1 | 1.6 |
| 1 | 4.9 | 3.1 | 1.5 | 0.1 | 2 | 5.4 | 3 | 4.5 | 1.5 |
| 1 | 5.4 | 3.7 | 1.5 | 0.2 | 2 | 6 | 3.4 | 4.5 | 1.6 |
| 1 | 4.8 | 3.4 | 1.6 | 0.2 | 2 | 6.7 | 3.1 | 4.7 | 1.5 |
| 1 | 4.8 | 3 | 1.4 | 0.1 | 2 | 6.3 | 2.3 | 4.4 | 1.3 |
| 1 | 4.3 | 3 | 1.1 | 0.1 | 2 | 5.6 | 3 | 4.1 | 1.3 |
| 1 | 5.8 | 4 | 1.2 | 0.2 | 2 | 5.5 | 2.5 | 4 | 1.3 |
| 1 | 5.7 | 4.4 | 1.5 | 0.4 | 2 | 5.5 | 2.6 | 4.4 | 1.2 |
| 1 | 5.4 | 3.9 | 1.3 | 0.4 | 2 | 6.1 | 3 | 4.6 | 1.4 |
| 1 | 5.1 | 3.5 | 1.4 | 0.3 | 2 | 5.8 | 2.6 | 4 | 1.2 |

Table 10.10: Iris Data (continued)

| Spec | Sepal Length | Sepal Width | Petal Length | Petal Width | Spec | Sepal Length | Sepal Width | Petal Length | Petal Width |
|---|---|---|---|---|---|---|---|---|---|
| 1 | 5.7 | 3.8 | 1.7 | 0.3 | 2 | 5 | 2.3 | 3.3 | 1 |
| 1 | 5.1 | 3.8 | 1.5 | 0.3 | 2 | 5.6 | 2.7 | 4.2 | 1.3 |
| 1 | 5.4 | 3.4 | 1.7 | 0.2 | 2 | 5.7 | 3 | 4.2 | 1.2 |
| 1 | 5.1 | 3.7 | 1.5 | 0.4 | 2 | 5.7 | 2.9 | 4.2 | 1.3 |
| 1 | 4.6 | 3.6 | 1 | 0.2 | 2 | 6.2 | 2.9 | 4.3 | 1.3 |
| 1 | 5.1 | 3.3 | 1.7 | 0.5 | 2 | 5.1 | 2.5 | 3 | 1.1 |
| 1 | 4.8 | 3.4 | 1.9 | 0.2 | 2 | 5.7 | 2.8 | 4.1 | 1.3 |
| 1 | 5 | 3 | 1.6 | 0.2 | 3 | 6.3 | 3.3 | 6 | 2.5 |
| 1 | 5 | 3.4 | 1.6 | 0.4 | 3 | 5.8 | 2.7 | 5.1 | 1.9 |
| 1 | 5.2 | 3.5 | 1.5 | 0.2 | 3 | 7.1 | 3 | 5.9 | 2.1 |
| 1 | 5.2 | 3.4 | 1.4 | 0.2 | 3 | 6.3 | 2.9 | 5.6 | 1.8 |
| 1 | 4.7 | 3.2 | 1.6 | 0.2 | 3 | 6.5 | 3 | 5.8 | 2.2 |
| 1 | 4.8 | 3.1 | 1.6 | 0.2 | 3 | 7.6 | 3 | 6.6 | 2.1 |
| 1 | 5.4 | 3.4 | 1.5 | 0.4 | 3 | 4.9 | 2.5 | 4.5 | 1.7 |
| 1 | 5.2 | 4.1 | 1.5 | 0.1 | 3 | 7.3 | 2.9 | 6.3 | 1.8 |
| 1 | 5.5 | 4.2 | 1.4 | 0.2 | 3 | 6.7 | 2.5 | 5.8 | 1.8 |
| 1 | 4.9 | 3.1 | 1.5 | 0.2 | 3 | 7.2 | 3.6 | 6.1 | 2.5 |
| 1 | 5 | 3.2 | 1.2 | 0.2 | 3 | 6.5 | 3.2 | 5.1 | 2 |
| 1 | 5.5 | 3.5 | 1.3 | 0.2 | 3 | 6.4 | 2.7 | 5.3 | 1.9 |
| 1 | 4.9 | 3.6 | 1.4 | 0.1 | 3 | 6.8 | 3 | 5.5 | 2.1 |
| 1 | 4.4 | 3 | 1.3 | 0.2 | 3 | 5.7 | 2.5 | 5 | 2 |
| 1 | 5.1 | 3.4 | 1.5 | 0.2 | 3 | 5.8 | 2.8 | 5.1 | 2.4 |
| 1 | 5 | 3.5 | 1.3 | 0.3 | 3 | 6.4 | 3.2 | 5.3 | 2.3 |
| 1 | 4.5 | 2.3 | 1.3 | 0.3 | 3 | 6.5 | 3 | 5.5 | 1.8 |
| 1 | 4.4 | 3.2 | 1.3 | 0.2 | 3 | 7.7 | 3.8 | 6.7 | 2.2 |
| 1 | 5 | 3.5 | 1.6 | 0.6 | 3 | 7.7 | 2.6 | 6.9 | 2.3 |
| 1 | 5.1 | 3.8 | 1.9 | 0.4 | 3 | 6 | 2.2 | 5 | 1.5 |
| 1 | 4.8 | 3 | 1.4 | 0.3 | 3 | 6.9 | 3.2 | 5.7 | 2.3 |
| 1 | 5.1 | 3.8 | 1.6 | 0.2 | 3 | 5.6 | 2.8 | 4.9 | 2 |
| 1 | 4.6 | 3.2 | 1.4 | 0.2 | 3 | 7.7 | 2.8 | 6.7 | 2 |
| 1 | 5.3 | 3.7 | 1.5 | 0.2 | 3 | 6.3 | 2.7 | 4.9 | 1.8 |
| 1 | 5 | 3.3 | 1.4 | 0.2 | 3 | 6.7 | 3.3 | 5.7 | 2.1 |
| 2 | 7 | 3.2 | 4.7 | 1.4 | 3 | 7.2 | 3.2 | 6 | 1.8 |
| 2 | 6.4 | 3.2 | 4.5 | 1.5 | 3 | 6.2 | 2.8 | 4.8 | 1.8 |
| 2 | 6.9 | 3.1 | 4.9 | 1.5 | 3 | 6.1 | 3 | 4.9 | 1.8 |
| 2 | 5.5 | 2.3 | 4 | 1.3 | 3 | 6.4 | 2.8 | 5.6 | 2.1 |
| 2 | 6.5 | 2.8 | 4.6 | 1.5 | 3 | 7.2 | 3 | 5.8 | 1.6 |
| 2 | 5.7 | 2.8 | 4.5 | 1.3 | 3 | 7.4 | 2.8 | 6.1 | 1.9 |

Table 10.10: Iris Data (continued)

| Spec | Sepal Length | Sepal Width | Petal Length | Petal Width | Spec | Sepal Length | Sepal Width | Petal Length | Petal Width |
|------|------|------|------|------|------|------|------|------|------|
| 2 | 6.3 | 3.3 | 4.7 | 1.6 | 3 | 7.9 | 3.8 | 6.4 | 2 |
| 2 | 4.9 | 2.4 | 3.3 | 1 | 3 | 6.4 | 2.8 | 5.6 | 2.2 |
| 2 | 6.6 | 2.9 | 4.6 | 1.3 | 3 | 6.3 | 2.8 | 5.1 | 1.5 |
| 2 | 5.2 | 2.7 | 3.9 | 1.4 | 3 | 6.1 | 2.6 | 5.6 | 1.4 |
| 2 | 5 | 2 | 3.5 | 1 | 3 | 7.7 | 3 | 6.1 | 2.3 |
| 2 | 5.9 | 3 | 4.2 | 1.5 | 3 | 6.3 | 3.4 | 5.6 | 2.4 |
| 2 | 6 | 2.2 | 4 | 1 | 3 | 6.4 | 3.1 | 5.5 | 1.8 |
| 2 | 6.1 | 2.9 | 4.7 | 1.4 | 3 | 6 | 3 | 4.8 | 1.8 |
| 2 | 5.6 | 2.9 | 3.6 | 1.3 | 3 | 6.9 | 3.1 | 5.4 | 2.1 |
| 2 | 6.7 | 3.1 | 4.4 | 1.4 | 3 | 6.7 | 3.1 | 5.6 | 2.4 |
| 2 | 5.6 | 3 | 4.5 | 1.5 | 3 | 6.9 | 3.1 | 5.1 | 2.3 |
| 2 | 5.8 | 2.7 | 4.1 | 1 | 3 | 5.8 | 2.7 | 5.1 | 1.9 |
| 2 | 6.2 | 2.2 | 4.5 | 1.5 | 3 | 6.8 | 3.2 | 5.9 | 2.3 |
| 2 | 5.6 | 2.5 | 3.9 | 1.1 | 3 | 6.7 | 3.3 | 5.7 | 2.5 |
| 2 | 5.9 | 3.2 | 4.8 | 1.8 | 3 | 6.7 | 3 | 5.2 | 2.3 |
| 2 | 6.1 | 2.8 | 4 | 1.3 | 3 | 6.3 | 2.5 | 5 | 1.9 |
| 2 | 6.3 | 2.5 | 4.9 | 1.5 | 3 | 6.5 | 3 | 5.2 | 2 |
| 2 | 6.1 | 2.8 | 4.7 | 1.2 | 3 | 6.2 | 3.4 | 5.4 | 2.3 |
| 2 | 6.4 | 2.9 | 4.3 | 1.3 | 3 | 5.9 | 3 | 5.1 | 1.8 |

## 10.7.1   Classification

The sepal length, sepal width, petal length, and petal width of 150 iris plants were recorded by Fisher [1936] and are reproduced in Table 10.10. Our first clues to the number of subpopulations or categories of iris, as well as to the general shape of the underlying frequency distribution, come from consideration of the histogram Figure 10.10. A glance suggests the presence of at least two species, though because of the overlap of the various subpopulations it is difficult to be sure. Three species actually are present as shown in Figure 10.8.

Proceding a variable at a time, CART$^{TM}$constructs a classification tree for the iris data consisting of two nodes. First, CART establishes that petal length is the most informative of the four variables, and the tree is split on the classification rule "petal length less than or equal to 2.450." Node 2 is split on the rule "petal width less than or equal to 1.750." CART concludes the remaining variables do not contribute significant additional information.

As shown in Table 10.11, 96% are correctly classified! A moment's thought dulls the excitement; this is the same set of observations we used to develop our selection criteria and of course we were successful. It's our success with new plants whose species we don't know in advance that is the true test.

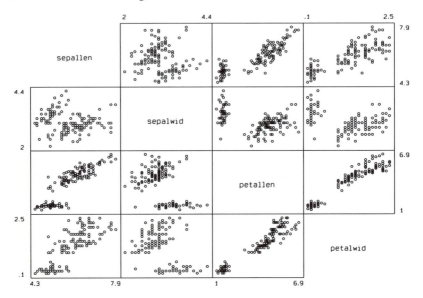

**Figure 10.10.**    Sepal and petal measurements of 150 iris plants.

Not having a set of yet-to-be-classified plants on hand, we make use of the existing data as follows: Divide the original group of 150 plants into 10 subgroups, each consisting of 15 plants. Apply the binary-tree-fitting procedure 10 times in succession, each time omitting one of the 10 groups, each time applying the tree it developed to the omitted group. Only 10 of the 150 plants or 7.5% were misclassified (Table 10.12), a 92.5% success rate, suggesting that in future samples consisting of irises of unknown species, the system of classification developed here would be successful 92.5% of the time.

### Refining the Discrimination Scheme

CART has three other features that command our attention. First, it is parsimonious; that is, all other factors being equal, it is programmed to produce a tree with the fewest number of branches. Only two of the four variables were utilized in the present example.

Second, CART may be programmed to take account of the expected composition of the target population. If we know we'll be searching in an area where species 3 comprises almost 50% of the population (in contrast to the training set in which species 3 is relatively rare), we can specify a high a priori probability for species 3 when setting up the tree-building procedure.

Suppose we are using CART to aid us in appraising a set of credit applications. We might have in mind three categories: i) grant credit, ii) request additional information, or 3) reject the application. Quite different costs are associated with each form of misclassification. We can specify these costs when setting up the

Table 10.11. Learning Sample Classification Table

| Actual | | Predicted Class | | | Actual |
|---|---|---|---|---|---|
| Class | | 1 | 2 | 3 | Total |
| | 1 | 50.000 | 0.000 | 0.000 | 50.000 |
| | 2 | 0.000 | 49.000 | 1.000 | 50.000 |
| | 3 | 0.000 | 5.000 | 45.000 | 50.000 |
| Predicted Total | | 50.000 | 54.000 | 46.000 | 150.000 |
| Correct | | 1.000 | 0.980 | 0.900 | |
| Success Total | | 0.667 | 0.647 | 0.567 | |
| Total Correct | | 0.960 | | | |

Table 10.12. Cross Validation Classification Table

| | | Cross Validation | | | Learning Sample | | |
|---|---|---|---|---|---|---|---|
| | Prior | | Mis- | | | Mis- | |
| Class | Probability | $N$ | classified | Cost | $N$ | classified | Cost |
| 1 | 0.333 | 50 | 0 | 0.000 | 50 | 0 | 0.000 |
| 2 | 0.333 | 50 | 5 | 0.100 | 50 | 1 | 0.020 |
| 3 | 0.333 | 50 | 5 | 0.100 | 50 | 5 | 0.100 |
| Tot | 1.000 | 150 | 10 | | 150 | 6 | |

tree-building procedure, ensuring their relative values will be accounted for in the discrimination rule.

## 10.7.2  Combined Approach

The best approach to model development may be to first use CART or some other data-mining tool to pinpoint essential variables and to segment the sample space based on the values of these variables. Then, apply traditional least-squares or LAD methods to find the optimal model and parameter values within each such segment.

A number of vendors have claimed success with just such an approach. At the time this text went to press, we were unable to confirm such claims or to obtain success at segmentation with the data sets considered in Section 10.6.2. Before purchasing any data mining software, we would advise testing and validating with your own data.

On the other hand, the technology of data mining is advancing rapidly. We have placed our estimation and validation data on our web site for use by the interested reader. Download from http://users.oco.net/drphilgood/data.zip.

## 10.8  Summary

In this chapter, you reviewed the steps in model development beginning with descriptive statistics such as scatter plots and correlations to identify relationships among variables. You learned a number of methods for ascertaining statistical significance. You learned it is essential to distinguish between goodness of fit and prediction, and were provided with a variety of cross-validation resampling methods to tie the two together. You were introduced to stepwise multiple regression and CART and saw how their use in combination might produce models of predictive value.

## 10.9   To Learn More

Regression is a rich and complex topic. Ryan [1997] provides a review of the many methods of estimation. Mosteller and Tukey [1977] document the many pitfalls. An excellent example of LAD regression is given in Cade and Richards [1996]. Good [2000; pp. 127–130] discusses the problems of exchangeability associated with multiple regression. Gallant [1987] and Ratkowsky [1990] provide introductions to nonlinear statistical models.

The resampling approach to model validation has been used in DNA sequencing (Doolittle [1981] and Karlin et al. [1983] adopt permutation methods while Hasegawa, Kishino, and Yano [1988] apply the bootstrap), ecology [Solow 1990], medicine [Arndt et al. 1996; Bullmore et al. 1996; Thompson et al. 1981; Titterington et al. 1981], nuclear power generation [Dubuisson and Lavison 1980], physics [Priesendorfer and Barnett 1983; DeJager, Swanepoel, and Raubenheimer 1986], archeology [Berry et al. 1980], business administration [Carter and Catlett 1987; Chhikara 1989; Thompson, Bridges, and Ensor 1992], and chemistry [Penninckz et al. 1996].

Mielke [1986] reviews alternate metrics. Marron [1987], Hjorth [1994], and Stone [1974] provide a survey of parameter selection and cross-validation methods. Techniques for model validation also are reviewed in Shao and Tu [1995; p. 306–313] who show that the delete–50% method, first proposed by Geisser [1975], is far superior to delete–1.

CART methodology is due to Breiman et al. [1984]. CART's use of cross-validation is studied by Breiman [1992] and Shao and Tu [1995].

## 10.10   Exercises

1. Characterize the following relationships as positive linear, negative linear, or nonlinear:
   a. Distance from the top of a bathtub to the surface of the water as a function of time when you fill the tub, then pull the plug.
   b. Same problem, only now you turn on the taps full blast.
   c. Sales as a function of your advertising budget.

**d.** Blood pressure as you increase the dose of a blood-pressure lowering medicine.

**e.** Electricity use as a function of the day of the year.

**f.** Number of bacteria at the site of infection as you increase the dose of an antibiotic.

**g.** Size of an untreated tumor over time.

2. Construct a scatter plot for the law school data in Exercise1 14, if you haven't already done so. What do you think is the nature of the association between undergraduate GPA and LSAT score?

3. Which is the cause and which the effect?
   **a.** overpopulation and poverty
   **b.** highway speed limits and number of accidents
   **c.** cases of typhus and water pollution

4. In Table 3.1, breaks appear to be related to log dose in accordance with the formula breaks $= a + b \log[\text{dose} + 0.01]$
   **a.** estimate $b$
   **b.** what does $a$ represent? How would you estimate $a$?

5. Reexamine the executive compensation data of Exercise 2 in Chapter 1. Do you feel there is a relationship between executive compensation and sales? executive compensation and return on equity? Can you encapsulate your gut feelings in model form?

6. Do you remember your own growth history? Bet your mother does. Plot a growth curve for yourself (your child, your grandchild) similar to that of Figure 10.3.

7. Suppose you wanted to predict the sales of automobile parts; which observations would be critical? Number of automobiles in service? Number of automobiles under warranty? Number of automobiles more than a given number of years in service? M2, the total money supply that is readily available in cash, savings and checking accounts, and so forth? Retail sales? Consumer confidence? Weather forecasts? Would your answer depend on whether you owned a dealership or an independent automobile parts chain? (By the way, in the studies Mark Kaiser and I completed, weather was the most significant variable. Go figure.)

8. Compute and compare the apparent error rates of the two Fawlty Towers prediction models.

9. Use the delete-50% method to compare the two Fawlty Towers prediction

models.

10. Suppose you've collected the prices of a number of stocks over a period of time and have developed a model with which to predict their future behavior. How would you go about validating your model?

11. Growth rates for chicks whose calorie-deficient diet was supplemented with vitamin B are summarized below by log-dose and sex. Is the dose response the same for the two sexes?

|  | 0.301 | 0.602 | 0.903 |
|---|---|---|---|
| males | 17.1, 14.3, 21.6 | 24.5, 20.6, 23.8 | 27.7, 31.0, 29.4 |
| females | 18.5, 22.1, 15.3 | 23.6, 26.9, 20.2 | 24.3, 27.1, 30.1 |
|  | 1.204 | 1.505 |  |
| males | 28.6, 34.2, 37.3 | 33.3, 31.8, 40.2 |  |
| females | 30.3, 33.0, 35.8 | 32.6, 36.1, 30.5 |  |

12. Homer and Lemeshow [1989] provide data on 189 births at a U.S. hospital. The last value in each row in the table below is the birth weight in grams. How would you go about predicting a low birth weight (defined as less than 2.5 kilograms)? Reading across each row the variables are observation number, low birth weight? (no 0, yes 1), age of mother in years, mother's weight in pounds at last menstrual period, race (white 1, black 2, other 3), smoked during pregnancy? (no 0, yes 1), number of previous premature labors, history of hypertension (no 0, yes 1), has uterine irritability (no 0, yes 1), number of physician visits in the first trimester, and birth weight in grams.

4,1,28,120,3,1,1,0,1,0,709

11,1,34,187,2,1,0,1,0,0,1135

15,1,25,85,3,0,0,0,1,0,1474

17,1,23,97,3,0,0,0,1,1,1588

19,1,24,132,3,0,0,1,0,0,1729

22,1,32,105,1,1,0,0,0,0,1818

24,1,25,115,3,0,0,0,0,0,1893

26,1,25,92,1,1,0,0,0,0,1928

28,1,21,200,2,0,0,0,1,2,1928

30,1,21,103,3,0,0,0,0,0,1970

32,1,25,89,3,0,2,0,0,1,2055

34,1,19,112,1,1,0,0,1,0,2084

36,1,24,138,1,0,0,0,0,0,2100

40,1,20,120,2,1,0,0,0,3,2126

43,1,27,130,2,0,0,0,1,0,2187

45,1,17,110,1,1,0,0,0,0,2225

47,1,20,109,3,0,0,0,0,0,2240

50,1,18,110,2,1,1,0,0,0,2296

10,1,29,130,1,0,0,0,1,2,1021

13,1,25,105,3,0,1,1,0,0,1330

16,1,27,150,3,0,0,0,0,0,1588

18,1,24,128,2,0,1,0,0,1,1701

20,1,21,165,1,1,0,1,0,1,1790

23,1,19,91,1,1,2,0,1,0,1885

25,1,16,130,3,0,0,0,0,1,1899

27,1,20,150,1,1,0,0,0,2,1928

29,1,24,155,1,1,1,0,0,0,1936

31,1,20,125,3,0,0,0,1,0,2055

33,1,19,102,1,0,0,0,0,2,2082

35,1,26,117,1,1,1,0,0,0,2084

37,1,17,130,3,1,1,0,1,0,2125

42,1,22,130,1,1,1,0,1,1,2187

44,1,20,80,3,1,0,0,1,0,2211

46,1,25,105,3,0,1,0,0,1,2240

49,1,18,148,3,0,0,0,0,0,2282

51,1,20,121,1,1,1,0,1,0,2296

52,1,21,100,3,0,1,0,0,4,2301

54,1,26,96,3,0,0,0,0,0,2325

56,1,31,102,1,1,1,0,0,1,2353

57,1,15,110,1,0,0,0,0,0,2353

59,1,23,187,2,1,0,0,0,1,2367

60,1,20,122,2,1,0,0,0,0,2381

61,1,24,105,2,1,0,0,0,0,2381

62,1,15,115,3,0,0,0,1,0,2381

63,1,23,120,3,0,0,0,0,0,2410

65,1,30,142,1,1,1,0,0,0,2410

67,1,22,130,1,1,0,0,0,1,2410

68,1,17,120,1,1,0,0,0,3,2414

69,1,23,110,1,1,1,0,0,0,2424

71,1,17,120,2,0,0,0,0,2,2438

 75,1,26,154,3,0,1,1,0,1,2442

76,1,20,105,3,0,0,0,0,3,2450

77,1,26,190,1,1,0,0,0,0,2466

78,1,14,101,3,1,1,0,0,0,2466

79,1,28,95,1,1,0,0,0,2,2466

81,1,14,100,3,0,0,0,0,2,2495

82,1,23,94,3,1,0,0,0,0,2495

83,1,17,142,2,0,0,1,0,0,2495

84,1,21,130,1,1,0,1,0,3,2495

85,0,19,182,2,0,0,0,1,0,2523

86,0,33,155,3,0,0,0,0,3,2551

87,0,20,105,1,1,0,0,0,1,2557

88,0,21,108,1,1,0,0,1,2,2594

89,0,18,107,1,1,0,0,1,0,2600

91,0,21,124,3,0,0,0,0,0,2622

92,0,22,118,1,0,0,0,0,1,2637

93,0,17,103,3,0,0,0,0,1,2637

94,0,29,123,1,1,0,0,0,1,2663

95,0,26,113,1,1,0,0,0,0,2665

96,0,19,95,3,0,0,0,0,0,2722

97,0,19,150,3,0,0,0,0,1,2733

98,0,22,95,3,0,0,1,0,0,2751

99,0,30,107,3,0,1,0,1,2,2750

100,0,18,100,1,1,0,0,0,0,2769

101,0,18,100,1,1,0,0,0,0,2769

102,0,15,98,2,0,0,0,0,0,2778

103,0,25,118,1,1,0,0,0,3,2782

104,0,20,120,3,0,0,0,1,0,2807

105,0,28,120,1,1,0,0,0,1,2821

106,0,32,121,3,0,0,0,0,2,2835

107,0,31,100,1,0,0,0,1,3,2835

108,0,36,202,1,0,0,0,0,1,2836

109,0,28,120,3,0,0,0,0,0,2863

111,0,25,120,3,0,0,0,1,2,2877

112,0,28,167,1,0,0,0,0,0,2877

113,0,17,122,1,1,0,0,0,0,2906

114,0,29,150,1,0,0,0,0,2,2920

115,0,26,168,2,1,0,0,0,0,2920

116,0,17,113,2,0,0,0,0,1,2920

117,0,17,113,2,0,0,0,0,1,2920

118,0,24,90,1,1,1,0,0,1,2948

119,0,35,121,2,1,1,0,0,1,2948

120,0,25,155,1,0,0,0,0,1,2977

121,0,25,125,2,0,0,0,0,0,2977

123,0,29,140,1,1,0,0,0,2,2977

124,0,19,138,1,1,0,0,0,2,2977

125,0,27,124,1,1,0,0,0,0,2922

126,0,31,215,1,1,0,0,0,2,3005

127,0,33,109,1,1,0,0,0,1,3033

128,0,21,185,2,1,0,0,0,2,3042

129,0,19,189,1,0,0,0,0,2,3062

130,0,23,130,2,0,0,0,0,1,3062

131,0,21,160,1,0,0,0,0,0,3062

132,0,18,90,1,1,0,0,1,0,3062

133,0,18,90,1,1,0,0,1,0,3062

34,0,32,132,1,0,0,0,0,4,3080

135,0,19,132,3,0,0,0,0,0,3090

136,0,24,115,1,0,0,0,0,2,3090

137,0,22,85,3,1,0,0,0,0,3090

138,0,22,120,1,0,0,1,0,1,3100

139,0,23,128,3,0,0,0,0,0,3104

140,0,22,130,1,1,0,0,0,0,3132

141,0,30,95,1,1,0,0,0,2,3147

42,0,19,115,3,0,0,0,0,0,3175

143,0,16,110,3,0,0,0,0,0,3175

144,0,21,110,3,1,0,0,1,0,3203

145,0,30,153,3,0,0,0,0,0,3203

146,0,20,103,3,0,0,0,0,0,3203

147,0,17,119,3,0,0,0,0,0,3225

148,0,17,119,3,0,0,0,0,0,3225

149,0,23,119,3,0,0,0,0,2,3232

150,0,24,110,3,0,0,0,0,0,3232

151,0,28,140,1,0,0,0,0,0,3234

154,0,26,133,3,1,2,0,0,0,3260

155,0,20,169,3,0,1,0,1,1,3274

56,0,24,115,3,0,0,0,0,2,3274

159,0,28,250,3,1,0,0,0,6,3303

160,0,20,141,1,0,2,0,1,1,3317

161,0,22,158,2,0,1,0,0,2,3317

162,0,22,112,1,1,2,0,0,0,3317

163,0,31,150,3,1,0,0,0,2,3321

164,0,23,115,3,1,0,0,0,1,3331

166,0,16,112,2,0,0,0,0,0,3374

167,0,16,135,1,1,0,0,0,0,3374

168,0,18,229,2,0,0,0,0,0,3402

169,0,25,140,1,0,0,0,0,1,3416

170,0,32,134,1,1,1,0,0,4,3430

172,0,20,121,2,1,0,0,0,0,3444

173,0,23,190,1,0,0,0,0,0,3459

174,0,22,131,1,0,0,0,0,1,3460

175,0,32,170,1,0,0,0,0,0,3473

176,0,30,110,3,0,0,0,0,0,3544

177,0,20,127,3,0,0,0,0,0,3487

179,0,23,123,3,0,0,0,0,0,3544

180,0,17,120,3,1,0,0,0,0,3572

181,0,19,105,3,0,0,0,0,0,3572

182,0,23,130,1,0,0,0,0,0,3586

183,0,36,175,1,0,0,0,0,0,3600

184,0,22,125,1,0,0,0,0,1,3614

185,0,24,133,1,0,0,0,0,0,3614

186,0,21,134,3,0,0,0,0,2,3629

187,0,19,235,1,1,0,1,0,0,3629

188,0,25,95,1,1,3,0,1,0,3637

189,0,16,135,1,1,0,0,0,0,3643

190,0,29,135,1,0,0,0,0,1,3651

191,0,29,154,1,0,0,0,0,1,3651

192,0,19,147,1,1,0,0,0,0,3651

193,0,19,147,1,1,0,0,0,0,3651

195,0,30,137,1,0,0,0,0,1,3699

196,0,24,110,1,0,0,0,0,1,3728

197,0,19,184,1,1,0,1,0,0,3756

199,0,24,110,3,0,1,0,0,0,3770

200,0,23,110,1,0,0,0,0,1,3770

201,0,20,120,3,0,0,0,0,0,3770

202,0,25,241,2,0,0,1,0,0,3790

203,0,30,112,1,0,0,0,0,1,3799

204,0,22,169,1,0,0,0,0,0,3827

205,0,18,120,1,1,0,0,0,2,3856

206,0,16,170,2,0,0,0,0,4,3860

207,0,32,186,1,0,0,0,0,2,3860

208,0,18,120,3,0,0,0,0,1,3884

209,0,29,130,1,1,0,0,0,2,3884

210,0,33,117,1,0,0,0,1,1,3912

211,0,20,170,1,1,0,0,0,0,3940

212,0,28,134,3,0,0,0,0,1,3941

213,0,14,135,1,0,0,0,0,0,3941

214,0,28,130,3,0,0,0,0,0,3969

215,0,25,120,1,0,0,0,0,2,3983

216,0,16,95,3,0,0,0,0,1,3997

217,0,20,158,1,0,0,0,0,1,3997

218,0,26,160,3,0,0,0,0,0,4054

219,0,21,115,1,0,0,0,0,1,4054

220,0,22,129,1,0,0,0,0,0,4111

221,0,25,130,1,0,0,0,0,2,4153

222,0,31,120,1,0,0,0,0,2,4167

223,0,35,170,1,0,1,0,0,1,4174

224,0,19,120,1,1,0,0,0,0,4238

225,0,24,116,1,0,0,0,0,1,4593

226,0,45,123,1,0,0,0,0,1,4990

# CHAPTER 11

# Which Statistic Should I Use?

This chapter provides you with an expert system for use in choosing an appropriate estimation or testing technique.

## 11.1  Parametric Versus Nonparametric

One should use a parametric test

- When you have a large number of observations ($> 40$) in each category; use Student's $t$, or the $F$ (see Chapter 4).

- When you have a very small number of observations ($< 6$) if the assumptions underlying the corresponding parametric test may be relied on (see Section 11.2).

Apart from these two exceptions, the resampling approach has several advantages over the parametric:

- The permutation test is exact under relatively nonstringent conditions. In the one-sample problem (which includes regression), observations must have symmetric distributions; in the 2- and $k$-sample problem, observations must be exchangeable among the samples.

- The permutation test provides protection against deviations from parametric assumptions, yet is usually as powerful as the corresponding unbiased parametric test even for small samples.

With two binomial or two Poisson populations, the most powerful unbiased permutation test and the most powerful parametric test coincide. With two normal populations, the most powerful unbiased permutation test and the most powerful unbiased parametric test produce identical results for very large samples.

Using a resampling method means you are no longer dependent on the availability of tables but can choose the statistic that is best adapted to your problem and to the alternatives of interest.

Consider a permutation test before you turn to a bootstrap. The bootstrap is not exact except for quite large samples and, often, is not very powerful. But as the bootstrap is essentially an estimation procedure, it can sometimes be applied when the permutation test fails: examples include the Behrens-Fisher problem, unbalanced designs, nonlinear regression, and tests of ratios and proportions.

## 11.2    But Is It a Normal Distribution?

To perform a parametric test, we must assume the observations come from a probability distribution that has a specific parametric form such as the binomial, Poisson, or Gaussian (normal) described in Section 2.2.

While there exist various formal statistical techniques for verifying whether a set of observations does or does not have a Poisson or normal distribution, the following informal guidelines are of value in practice:

- An observation has the Poisson distribution if it is the cumulative result of a large number of opportunities each of which has only a small chance of occurring. For example, if we seed a small number of cells into a Petri dish that is divided into a large number of squares, the distribution of cells per square follows the Poisson.

- An observation has the Gaussian or normal distribution if it is the sum of a large number of factors each of which makes a very small contribution to the total. This explains why the mean of a large number $N$ of observations, $\frac{1}{N}\sum_{i=1}^{N} X_i = \sum_{i=1}^{N} X_i/N$ will be normally distributed, even if the individual observations come from non-Gaussian distributions.

In many applications in economics and pharmacology where changes are best expressed in percentages, a variable may be the product of a large number of variables each of which makes only a very small contribution to the total. Such a variable has the lognormal distribution; as the logarithm of a product is the sum of the logarithms of the individual terms as in

$$\log[X_1 X_2 \cdots X_n] = \sum \log[X_i],$$

its logarithm has a normal distribution.

## 11.3    Which Hypothesis?

Recall that with a permutation test, we

1. Choose a test statistic $S(X)$.

2. Compute $S$ for the original set of observations. (We may first transform to ranks to minimize the effects of a few extreme observations.)

3. Obtain the resampling distribution of $S$ by repeatedly rearranging the data. With two or more samples, we combine all the observations into a single large sample before we rearrange them.

4. Accept or reject the null hypothesis according to whether $S$ for the original observations is smaller or larger than the upper-percentage point of the resampling distribution.

To obtain a nonparametric bootstrap, we

1. Choose a test statistic $S(X)$.

2. Compute $S$ for the original set of observations.

3. Obtain the bootstrap distribution of $S$ by repeatedly resampling from the observations. We need not combine the samples, but may resample separately from each sample. We resample with replacement. (Afterward we may smooth, correct for bias, accelerate, or iterate to obtain a more accurate resampling distribution.)

4. Accept or reject the null hypothesis according to whether $S$ for the original observations is smaller or larger than the upper-percentage point of the (revised) resampling distribution.

To obtain a parametric test, we

1. Choose a test statistic $S$ whose distribution $F$ may be computed and tabulated independent of the observations.

2. Compute $S$ for the observations $X$. (We may use an initial transformation to make the various samples more alike in dispersion.)

3. Accept or reject the null hypothesis according to whether $S(X)$ is smaller or larger than the upper-percentage point of $F$.

In almost all the examples we've considered so far, the statistics $S$ used in the parametric, permutation, and bootstrap methods were similar, the difference lay in the methods by which critical values were determined. The nature of the hypothesis, the form of the test statistic, and the method all depend upon the approach we employ. For example,

*Parametric*: To test the hypothesis that two populations are normal with mean 0 and variance $\sigma^2$ against the alternative that the second population, while still normal with mean 0, has variance $\sigma_1^2 > \sigma^2$ the best test is a parametric one based on the distribution of the ratio of the sample variances.

*Permutation*: To test the hypothesis that two populations have the same distribution $F$ against the alternative that one population has the distribution $G$ with scale parameter such that $G[x/\sigma] = F[x]$, the best test is a permutation test based on the resampling distribution of the test statistic described in Section 3.3.

*Bootstrap*: To test the hypothesis that two populations have the same scale parameter $\sigma$ against the alternative that the second population has the scale parameter $\sigma_1 > \sigma$, the best test is based on the bootstrap estimate of the ratio of the two parameters.

## 11.4    A Guide to Selection

The initial division of this guide is into three groupings: categorical data, ordinal, nonmetric data, and metric data.

### 11.4.1    Data in Categories

Examples include men versus women, white versus black versus Hispanic versus other, and improved versus no change versus worse.

Only two factors are involved. For example, sex versus political party.

Each factor is at exactly two levels.

There is a single table: Use Fisher's exact test, Section 7.1.

There are several $2 \times 2$ tables: See Section 7.2.1.

One factor is at three or more levels.

This factor is not ordered (e.g., race).

You want a test that provides protection against a broad variety of alternatives: Use the permutation distribution of the chi-square statistic, Section 7.4.

You wish to test against the alternative of a cause-effect dependence: Use the Freeman and Halton test, Section 7.4.

This factor can be ordered: Assign scores to this factor based on your best understanding of its effects on the second variable and use Pitman correlation, Section 7.5.1.

Both factors are at three or more levels.

Neither factor can be ordered.

One factor might be caused or affected by the other: Use permutation distribution of Kendall's or Cochran's Q, Section 7.4.

Cause and effect relationship not suspected: Use permutation distribution of statistic, Section 7.4.

One factor can be ordered: Assign scores to this factor based on your best understanding of its effects on the second variable, or use the ANOVA with arbitrary scores, Section 7.5.2.

Both factors can be ordered:

Use Mantel's $U$, Section 9.3.1.

A third covariate factor is present: Use the method of Bross [1964].

## 11.4.2 Ordered Observations

The data are discrete, taking values in an ordered set such as {low, 1, 2, hi}.

Each sample consists of a fixed number of independent identically distributed observations that can be either 0 or 1. (A set of trials each of which may result in a success or a failure is an example.)

One or two samples: Use the parametric test for the binomial. See for example, Lehmann [1986, pp. 81, 154]. A confidence interval for the binomial parameter may be obtained using the method of Section 5.2.1.

More than two samples, but only one factor: Analyze as indicated above under categorical data. More than one factor. Transform the data to equalize variances: For each factor combination, take the arcsin of the square root of the proportion of observations that take a nonzero value. Analyze as in Section 8.2.1.

Each sample consists of a set of independent identically distributed Poisson observations. One or two samples: Use the parametric test for the Poisson. In the two-sample case, the UMPU test uses the binomial distribution, Section 4.2.

More than two samples: Transform the data to equalize the variances by taking the square root of each observation. Analyze as in Section 8.2.1.

Each sample consists of a set of exchangeable observations whose distribution is unknown: If you can define a metric (see Section 10.4.1), do so and treat as metric data, otherwise treat as categorical data. In the former case, be cautious in interpreting a negative finding; the significance level may be large simply because the test statistic can take on only a few distinct values. See Section 3.4.1.

## 11.4.3 Continuous Data

If you recognize the data have the normal distribution, a parametric test like Student's $t$ or the $F$-ratio may be applicable. But you can protect yourself against deviations from normality by making use of the resampling distribution of these statistics rather than the tabulated distribution.

### Single Sample

You want to estimate the population median: Use the sample median; for a confidence interval, see Section 5.2.2.

You want to estimate the mean, standard deviation, correlation coefficient, or some other population parameter: Use the corresponding plug-in estimate and apply the methods of Chapter 5 to obtain a bootstrap confidence interval.

You want to test that the location parameter has a specific value. And feel safe in assuming the underlying distribution is symmetric about the location parameter: Use the permutation test described in Section 3.6.2.

If the distribution is not symmetric, but has a known parametric form, apply the corresponding parametric test as in Section 4.4.

If it does not have a known parametric form, Consider applying an initial transformation that will symmetrize the data. For example, take the logarithm of data that undergo percent age changes. Be aware such a transformation affects the form of the loss function.

Else bootstrap as in Section 3.7.1 and apply the refinements described in Sections 5.4.

You want to test that the scale parameter has a specific value.

First, divide each observation by the hypothesized value of the scale parameter. Then, apply one of the procedures noted above for testing a location parameter.

You want to assess the effects of covariates: See Section 5.3 and 10.2.3.

## Two Samples

You want to test whether the scale parameters are equal.

You know the means/medians of the two populations or you know they are equal: Use the permutation distribution of the ratio of the sample variances.

You have no information about the means/medians: Use Aly's test, Section 3.3.

You want to test whether the location parameters of the two populations are equal:

If changes are proportional rather than additive, use the logarithms of the observations.

If you suspect outliers, use ranks.

If the data have been Type I censored, see Good [2000, p. 167].

Each sample consists of measures taken on different subjects.

Use two-sample comparison described in Section 3.1. For a confidence interval for the difference in location parameters, see Section 5.2.2. For a confidence interval for the ratio of two parameters, use the bootstrap, see Sections 5.2 and 5.4.

Two observations were made on each subject; these observations are to be compared: Use the matched-pair comparison described in Section 3.7.

To obtain a confidence interval for the difference, see Section 5.2.2.

To compare the slopes of two regression lines, see Section 10.2.4.

## More Than Two Samples

If changes are proportional rather than additive, work with logarithms of the observations.

If you can take advantage of other factors to block the samples, then rerandomize on a block-by-block basis.

A single factor distinguishes the various samples (see Sections 8.1 and 8.2).

Factor levels not ordered: Use permutation distribution of an $F$-ratio.

Factor levels are ordered: Use Pitman's correlation.

Multiple factors are involved:

One of the factors consists of repeated measurements made over time: See Section 9.1.2.

Experimental design is balanced: Use one of the permutation techniques described in Section 8.2.

Experimental design not balanced: Use bootstrap as in Section 8.6.

# Appendix 1

# Program Your Own Resampling Statistics

Whether you program from scratch in BASIC, C, or FORTRAN, or use a macro-language like that provided by Stata, S-Plus, or SAS, you'll follow the same four basic steps:

1. Read in the observations.

2. Compute the test statistic for these observations.

3. Resample repeatedly.

4. Compare the test statistic with its resampling distribution.

Begin by reading in the observations and, even if they form an array, store them in the form of a linear vector for ease in resampling. At the same time, read and store the sample sizes so the samples can be recreated from the vector. For example, if we have two samples 1, 2, 3, 4 and 1, 3, 5, 6, 7, we would store the observations in one vector $X = (1\ 2\ 3\ 4\ 1\ 3\ 5\ 6\ 7)$ and the sample sizes in another (4 5) so as to be able to mark where the first sample ends and the next begins.

Compute the value of the test statistic for the data you just read in. If you are deriving a permutation statistic, several short cuts can help you here. Consider the well-known $t$-statistic for comparing the means of two samples:

$$t = \frac{\bar{X} - \bar{Y}}{\sqrt{S_x^2 + S_y^2}}.$$

The denominator is the same for all permuations, so don't wast time computing it, instead use the statistic

$$t' = \bar{X} - \bar{Y}.$$

A little algebra shows that

$$t' = \sum X_i \left(\frac{1}{n_x} + \frac{1}{n_y}\right) - \left(\sum X_i + \sum Y_i\right)\frac{1}{n_y}.$$

But the sum $(\sum X_i + \sum Y_i)$ of all the observations is the same for all permutations, as are the sample sizes $n_x$ and $n_y$; eliminating these terms (why do extra extra work if you don't have to?) leaves

$$t'' = \sum X_i$$

the sum of the observations in the first sample, a sum that is readily computed.

*Whatever you choose as your test statistic, eliminate terms that are the same for all permutations before you begin.*

---

**Four Steps to a Resampling Statistic**

1. Get the Data.
2. Compute the Statistic $S_0$ for the Original Observations.
3. Loop:

   a. Resample.
   b. Compute the statistic $S$ for the resample.
   c. Determine whether $S > S_0$.

4. State the $p$-value.

---

How we resample will depend on whether we use the bootstrap or a permutation test. For the bootstrap, we repeatedly sample with replacement from the original samplefor the permuation test, without. Some examples of program code are given in a sidebar. More extensive examples entailing corrections for bias and acceleration are given in Appendix 3.

---

**Computing a Bootstrap Estimate**

**C++ Code**

```
#include <stdlib.h>
get_data(); //put the variable of interest in the first n
            //elements of the array X[].
randomize(); //initializes random number generator
for (i=0; i<100; i++){
  for (j=0; j<n; j++) Y[j]=X[random(n)];
  Z[i]=compute_statistic(Y);
  //compute the statistic for the array Y and store
  //it in Z compute_stats(Z);
```

---

### Computing a Bootstrap Estimate

## GAUSS

```
n = rows(Y);
U = rnds(n,1, integer seed));
I = trunc(n*U + ones(n,1,1));
    Ystar = Y[I,.];
```

(This routine only generates a single resample.)

## SAS Code

```
PROC IML
  n = nrow(Y);
  U = ranuni(J(N,1, integer seed));
  I = int(n*U + J(n,1,1));
      Ystar = Y(|I,|);
```

## S-Plus Code

```
"bootstrap" <- function(x, nboot, statistic,. . .){
data <- matrix(sample(x,size=length(x)*nboot, replace=T),
                        nrow=nboot)
return(apply(data,1,statistic,. . .))
}
# where statistic is an S+ function which calculates the
#statistic of interest.
```

## Stata Code

```
. use data, clear
. bstrap median, args(height), reps(100)
  where median has been predefined as follows:
program define median
        if '''1''' == ''?'' {
                global S_1 "median"
                exit
                }
summarize '2',detail
post '1' _result(10)
end
```

---

## IMPORTANCE SAMPLING

We may sample so as to give equal probability to each of the observations or we may use importance sampling in which we give greater weight to some of the observations and less weight to others.

The idea behind importance sampling is to reduce the variance of the estimate and, thus, the number of resamples necessary to achieve a given level of confidence.

Suppose we are testing the hypothesis that the population mean is 0 and our original sample contains the observations $-2$, $-1$, $-0.5$, 0, 2, 3, 3.5, 4, 7, and 8. Resamples containing 7 and 8 are much more likely to have large means, so instead of drawing bootstrap samples such that every observation has the same probability 1/10 of being included, we'll weight the sample so that larger values are more likely to be drawn, selecting $-2$ with probability 1/55, $-1$ with probability 2/55, and so forth, with the probability of selecting 8 being 9/55th's.[1] Let $I(S^* > 0) = 1$ if the mean of the bootstrap sample is greater than 0 and 0 otherwise. The standard estimate of the proportion of bootstrap samples greater than zero is given by

$$\frac{1}{B} \sum_{b=1}^{B} I(S_b^* > 0).$$

Because since we have selected the elements of our bootstrap samples with unequal probability, we have to use a weighted estimate of the proportion

$$\frac{1}{B} \sum_{b=1}^{B} I(S_b^* > 0) \frac{\prod_{i=1}^{10} \pi_i^{t_{ib}}}{.1^{10}},$$

where $\pi_i$ is the probability of selecting the $i$th element of the original sample and $t_{ib}$ is the number of times the $i$th element appears in the $b$th bootstrap sample. Efron and Tibshirani [1993, p. 355] found a seven-fold reduction in variance using importance sampling with samples of size 10.

Importance sampling for bootstrap tail probabilities is discussed in Johns [1988], Hinkley and Shi [1989], and Do and Hall [1991]. Mehta, Patel, and Senchaudhuri [1988] considered importance sampling for permutation tests.

## REARRANGEMENTS

For permutation tests, three alternative approaches to resampling suggest themselves:

- Exhaustive—examine all possible resamples.
- Focus on the tails—examine only samples more extreme than the original.
- Monte Carlo.

---

[1] $1 + 2 + 3 + 4 + 5 + 6 + 7 + 8 + 9 + 10 = 55$.

---

### Estimating Permutation Test Significance Levels

**C++ Code for One-Sided Test:**

```cpp
int Nmonte; // number of Monte Carlo simulations
float stat0; // value of test statistic for the unpermuted
             //observations
float X[]; // vector containing the data
int n[]; // vector containing the cell/sample sizes
int N; // total number of observation

proc main() {
  int i, s, k, seed, fcnt=0;
  get_data(X,n);
  stat_orig = statistic(X,n); //compute statistic for the
                              //original sample
  srand(seed);

  for (i=0; i < Monte; i++){
    for (s = N; s > n[1]; s--)
      {
      k = choose(s); // choose a random integer from 0 to s-1
      temp = X[k];
      X[k] = X[s-1]; // put the one you chose at end of
                     //vector
      X[s-1] = temp;
      }

    if (stat_orig >= statistic(X,n)) fcnt++;
    //count only if the value of the statistic for the
    //rearranged sample is as or more extreme than its value
    //for the original sample.
    }
  cout << ``alpha = '' << fcnt/NMonte;
}

void get_data(float X, int n){
// This user-written procedure gets all the data and packs
//it into a single long linear vector X. The vector n is
//packed with the sample sizes.
}

float statistic(float X, int n){
// This user-written procedure computes and returns the
//test statistic.
{
```

The first two of these alternatives are highly-application specific. Zimmerman [1985a, b] provides algorithms for exhaustive enumeration of permutations. The computationally efficient algorithms developed in Mehta and Patel [1983] and Mehta, Patel, and Gray [1985] underlie the StatXact results for categorical and ordered data reported in in Chapter 7. Algorithms for exact inference in logistic regression are provided in Hirije, Mehta, and Patel [1988].

An accompanying sidebar provides the C++ code for estimating permutation test significance levels via a Monte Carlo. Unlike the bootstrap, permutations require we select elements with replacement. The trick (see sidebar) is to stick each selected observation at the end of the vector where it will not be selected a second time. Appendix 2 contains several specific application of this programming approach.

---

**Computing a Single Random Rearrangement**

With SAS:

```
PROC IML
n = nrow (Y);
U = randuni (J(n,1,seed));
I = rank (U);
Ystar = Y(|I,|);
```

With GAUSS

```
n = nrows (Y);
U = rnds (n,1,seed);
I = rankindx (U,1);
Ystar = Y[|I,|];
```

---

# Appendix 2

# C++, SC, and Stata Code for Permutation Tests

Simple variations of the C++ program provided in Appendix 1 yield many important test statistics.

## Estimating Significance Level of the Main Effect of Fertilizer on Crop Yield in a Balanced Design

Set aside space for

| | |
|---|---|
| Monte | the number of Monte Carlo simulations |
| $S_0$ | the original value of test statistic |
| $S$ | test statistic for rearranged data |
| data | {5, 10, 8, 15, 22, 18, 21, 29, 25, 6, 9, 12, 25, 32, 40, 55, 60, 48}; |
| $n = 3$ | number of observations |
| in each category blocks = 2 | number of blocks |
| levels = 3 | number of levels of factor |

```
MAIN program
GET_DATA
  put all the observations into a single linear vector
COMPUTE S_0 for the original observations
Repeat Monte times:
  for each block
    REARRANGE the data in the block
  COMPUTE S
  Compare S with S_0
PRINT out the proportion of times S was larger than S_0

REARRANGE
Set s to the number of observations in the block
Start: Choose a random integer k from 0 to s-1
  Swap X[k] and X[s-1]:
```

```
    Decrement s and repeat from start
Stop after you've selected all but one of the samples.

GET_DATA
user-written procedure gets data and packs it into a
two-dimensional array in which each row corresponds to
a block.

COMPUTE
F_1=sum_{i=1}^2 sum_{j=1}^3|X_{ij.}-X_{.j.}|
for each block
  calculate the mean of that block
  for each level within a block
    calculate the mean of that block-level
    calculate difference from block mean
```

## *Estimating Significance Level of the Interaction of Sunlight and Fertilizer on Crop Yield*

Test statistic based on the deviates from the additive model. Design must be balanced.

Set aside space for

| | |
|---|---|
| Monte | the number of Monte Carlo simulations |
| $S_0$ | the original value of test statistic |
| $S$ | test statistic for rearranged data |
| data | $\{5, 10, 8, 15, 22, 18, 21, 29, 25, 6, 9, 12, 25, 32, 40, 55, 60, 48\}$; |
| deviates | vector of deviates |
| $n = 3$ | number of observations in each category |
| blocks = 2 | number of blocks |
| levels = 3 | number of levels of factor |

```
MAIN program
GET_DATA
Calculate the DEVIATES
COMPUTE the test statistic S_0
Repeat Monte times:
    REARRANGE the observations
  COMPUTE the test statistic S
  Compare S with S_0
Print out the proportion of times S was larger than S_0

COMPUTE sum_i^I sum_j^J (sum_k^K X'_{ik})^2
  for each block
    for each level
      sum the deviates
      square this sum
      cumulate
```

```
DEVIATES X'_{ijk} = X_{ijk}-X_{i..} - X_{.j.} + X_{...}
Set aside space for level means, block means, and grand mean
for each level calculate mean
for each block
  calculate mean
  for each level
    cumulate grand mean
for each block
  for each level
    calculate deviate from additive model
```

## *Bounds on Significance Level of Type I Censored Data*

Set aside space for

| | |
|---|---|
| Monte | the number of Monte Carlo simulations |
| $T_0 = S_U(0) + C N_c(0)$ | the original value of the test statistic |
| $T = S_U(\pi) + C N_C(\pi)$ | test statistic for rearranged data |
| data | $\{880, 3430, 2860.C, 4750, 3820, C, 7170, C, 4800\}$; |
| $n = 5, m = 5$ | number of observations in each category |
| $C = 10,000$ | |

```
Main program:
GET_DATA
  put all the observations into a single linear vector
COMPUTE T_o for the original observations
Repeat Monte times:
  REARRANGE the data in the block
  COMPUTE S_U, N_C, T
  Compare T with T_0
PRINT out the proportion of times T was larger than T_0

REARRANGE
Start: Set s=n
  Choose a random integer k from 0 to s-1
  Swap X[k] and X[s-1]:
    Decrement s and repeat from start
  Stop when s=0.

GET_DATA
user-written procedure gets data and packs it into a
linear vector

COMPUTE
N_C = S_U =0
For i=1 to n
  If X[i] = C then N_C++
    else S_U + = X[i]
T = S_U + CN_C
```

## Two-Sample Test for Dispersion Using Aly's Statistic Written in $SC^{TM}$

```
good_aly(X,Y,s)
  X,Y -> data vectors, size nx,ny
  s-> no. of assignments to sample (default: all)
```

If $nx = ny$, $m = nx$, otherwise let $m$ be the smaller of $nx$, $ny$ and replace the larger vector by a sample of size $m$ from it. Let $x$, $y$ be the corresponding order statistics. Let $\{D_x, D_y\}$ be the $m - 1$ differences $x[i+1] - x[i]$ and $y[i+1] - y[i]$, $i = 1, \ldots, m - 1$, and let $w[i] = i(m - i)$. These $\{D_x, D_y\}$ are passed to perm2_(), which computes $w.\pi(D_x)$ where $\pi(D_x)$ is a random rearrangement of $m$ elements from $\{D_x, D_y\}$ and finds the two-tailed fraction of rearrangements for which $|w.D_x| \geq |w.\pi(D_x)|$.

```
func good_aly_f(){
  sort ($1$)
  return g_c*$i$
}

proc good_aly(){
  local(ac,el,obs,X,Y,m,mm1,m1)

  # check arguments
  if (ARGCHECK) {
    m1= ''(vec,vec) or (vec,vec,expr)\n''
    if(NARGS<2 || NARGS>3) abort_(ui,$0$,m1)
    if(!isvec($1$) || !isvec($2$)) abort_(ui,$0$,m1)
    if (NARGS>=3){
      if($3$!=nint($3$) || $3$<10) abort_($0$,a3od)
    }
  }

  # set 1-up element-numbering
  el= ELOFFSET; ELOFFSET= -1
  # arrive at, if necessary, equal-size (m) vectors
  X= $1$; Y= $2$
  if(sizeof(X)<sizeof(Y)){
    shuffle(Y); limit(Y,1,m=sizeof(X))
  }else if(sizeof(Y)<sizeof(X)){
    shuffle(X); limit(X,1,m=sizeof(Y))
  }else m= sizeof(X)
  mm1= m - 1

  # get the pairwise differences
  sort(X); X= X[2:m]-X[1:mm1]
  sort(Y); Y= Y[2:m]-Y[1:mm1]
  #get the multipliers for the differences, as global g_c
```

```
  vector(g_c ,mm1)
  ords (g_c)
  g_c= g_c *_(-g_c+m)
  # Result for the actual data:-
  print(''\tAly delta='',obs=(good_aly_f(Y) -
good_aly_f (X))/(m*m))


#Call perm2_() to give the left and right tail over all
  # (3 arguments) or some (4 arguments) using the statistic
  # returned by good_aly_f(): order or arguments is reversed
  # in perm2_() calls to make scale(Y) > scale(X) correspond
  # to a small upper tail probability, for consistency with
  # other similar sc routines

  # turn off argument-checking for the call to perm2_()
  ac= ARGCHECK; ARGCHECK= 0

  if(NARGS==2){
    # complete enumeration
    perm2_ good_aly_f,Y,X
  }else{
    # subset only
    perm2_(good_aly_f,Y,X,$3$)
  }

  # free globals
  free(g_c)

  # restore re-set globals
  AUX[5]= obs
  ARGCHECK= ac
  ELOFFSET= el
}
```

## Using Stata for Permutation Tests

A set of ado files for performing one-sample, two-sample, and k-sample tests
of location parameters created by Roberto G. Guiterrez of the Stata corporation
may be downloaded from http://users.oco.net/resamp.htm. A help file
and a test for correlation are included. Place the downloaded files in your \ado
directory (you may have to create this directory) or your Stata working directory.
Do *not* put them in the \stata\ado directory.

## Using Xlisp for the Bootstrap

To get the code you need, first go to http://www.stat.ucla.edu/develop/

`lisp/xlisp/xlisp-stat/code/`. Then go to statistics/simulation/bootstrap. This has the Efron and Tibshirani routines in Xlisp, but also the wild bootstrap, the balanced bootstrap, various regression bootstraps, bootstraps for dependent data, complex sampling bootstraps, and permutation tests. Of course you need Xlisp-Stat, but you needed that anyway.

## *To Learn More*

Consult Good [2000].

# Appendix 3

# Resampling Software

You can download evaluation copies of almost all the software listed below from the websites indicated in the text.

**DBMS Copy**, though not a statistics package, is a must for anyone contemplating the use of resampling methods. DBMS Copy lets you exchange files among two dozen statistics packages, a dozen plus data base managers, and multiple versions of a half dozen spreadsheets. This book could not have been put together without its aid. UNIX and Windows versions. Conceptual Software, 9660 Hillcroft #510, Houston, TX 77096, 800/328-2686. www.conceptual.com

**Biostat Software**, an eclectic set of executable algorithms, SAS macros, and FORTRAN source code, includes bootstrap estimates of the survival function using the Cox regression method, Kaplan-Meier estimate of the survivor function, model parameters and cell probabilities in logistic regression, and bootstrap bumping and double bumping. http://www.biostatsoftware.com

**Blossom Statistical Analysis Package** This interactive program for analyzing data utilizing multi-response permutation procedures (MRPP) includes statistical procedures for grouped data, agreement of model predictions, circular-distributions, goodness of fit, least absolute deviation and quantile regression. Programmed by Brian Cade at the U.S. Geological Survey, Midcontinent Ecological Science Center. PC only with online manual in HTML Frames format. Freeware. www.mesc.usgs.gov/blossom/blossom.html

**CART** provides all the flexibility of the CART methodology for discrimination. including prior probabilities and relative costs. Windows. A companion product **MARS** is said to automatically identify relevant predictors, transform variables to minimize residuals, determine interactions, and perform piecewise regression. Salford Systems, 5952 Bernadette Lane, San Diego, CA 92120. 619/543-8880. www.salford-systems.com

**NoName** lacks the pull-down menus and prompts that Windows users have grown to rely on. But it can be used to analyze all the permutation test and many of the simpler bootstraps described in this text. Best of all for the nonstatistician, it tries to guide you to the correct test. Shareware. (DOS only). May be downloaded from `http://users.oco.net/drphilgood`

   *   **NPC TEST** is said to provide for multi-aspect testing; analyses with missing values, repeated measures, and the analysis of experimental designs using synchronized permutations. A demonstration version, SAS macro, and S-Plus code may be downloaded from `www.stat.unipd.it/~pesarin/software.html`.

    **NPSTAT** carries out parametric and randomization tests on single factor designs, repeated measures, correlations, and Fisher's exact test. **NPFACT** support 2- and 3-factor designs. DOS. Freeware. `cgi.rmci.net/rmay/npstat.html`

    **nQuery Advisor** helps you determine sample size for 50+ design and analysis combinations. A challenge to use but the advice is excellent. Windows. Statistical Solutions, 8 South Bank, Crosse's Green, Cork, Ireland. 800/262-1171. +353 21 4319629. `www.statsol.ie`

    **PolyAnalyst** claims to automatically derive the best-fitting linear or nonlinear model under Windows NT. Also provides for neural net and GMDH hybrid, cluster detection, decision tree analysis, fuzzy logic classification, multiple group classification, market basket analysis, and stepwise regression. Windows. Megaputer Intelligence. 120 W. Seventh St., Suite 310, Bloomington, IN 47404. 812/330-0110. `www.megaputer.com`

   *   **Resampling Stats** is the beginner's best friend. It enables you to quickly generate test data sets and visualize the variation inherent in almost any real-world study. Available in several versions including an add-in for use with Excel. Windows. Resampling Stats, 612 N. Jackson St., Arlington, VA 22201. 703/522-2713. `www.resample.com`.

    **RT** is based on the programs in Brian Manly's book *Randomization and Monte Carlo Methods in Biology*. Analysis of variance with one to three factors, with randomization of observations or residuals, simple and linear multiple regression, with randomization of observations or residuals, Mantel's test with up to nine matrices of dependent variables, Mead's test and Monte Carlo tests for spatial patterns, tests for trend, autocorrelation and periodicity in time series. Particularly valuable for block matrix classification. Windows. Western EcoSystem Technology Inc., 2003 Central Ave., Cheyenne, WY 82007. 307/634-1756. `www.west-inc.com`

    **SAS** offers StatXact (less the manual) as an add-on. The statistical literature is filled with SAS macros for deriving bootstrap and density estimates. Offers a wider choice of ready-to-go graphics options than any other statistics package. SAS Institute Inc., SAS Campus Drive, Cary, NC 27513. (Windows, MVS, CMS, VMS, Unix). `www.sas.com`

    **SCRT** (Single Case Randomization Tests). Alternating Treatments Designs, AB Designs and extensions (ABA, ABAB, etc.), and Multiple Baseline Designs. Customized statistics may be built and selected as the criterion for the randomization test. Provides a nonparametric meta-analytic procedure. DOS only. Interuniversity Expertise Center ProGAMMA, P.O. Box 841, 9700 AV Groningen, the Netherlands.

    **Simstat** helps you derive the empirical sampling distribution of 27 different estimators. Provides for rank tests, measures of correlation, linear and nonlinear regression. Provalis Research, 2414 Bennett St, Montreal, QC, CANADA H1V 3S4. www.simstat.com

**S-PLUS** is a language, a programming environment, and a comprehensive package of subroutines for statistical and mathematical analysis. A dozen texts and hundreds of file repositories on the Internet now offer all the subroutines you'll need for bootstrap, density estimation, and graphics. For UNIX and Windows. Statistical Sciences, 1700 Westlake Avenue North, Suite 500, Seattle, WA 98109. (800) 569-0123. www:http://statsci.com

**S-PLUS** code for obtaining BCa intervals may be obtained from the statistics archive of Carnegie-Mellon University; by sending an email to statlib@lib. stat.cmu with the one-line message "snd bootstrap.funs from S". Hints on usage will be found in Efron and Tibshirani [1993].

**SPSS** offers ease of use, extensive graphics, and add-on modules for Exact Tests and Neural Nets. A bootstrap subcommand provides bootstrap estimates of the standard errors and confidence limits for constrained nonlinear regression. (Windows). SPSS Inc., 444 North Michigan Avenue, Chicago, Illinois 60611. 312/329-2400. www.spss.com

**Stata** provides a complete set of graphic routines plus subroutines and pre-programmed macros for bootstrap, density estimation, and permutation tests. (Windows, Unix.) Stata Corp, 702 University Drive East, College Station, TX 77840. 800/782-8272. www.stata.com An online course in its use in resampling may be obtained from www.timberlake.uk.

**Statistical Calculator** (SC) (DOS, Unix, T800 transputer.) An extensible statistical environment, supplied with over 1200 built-in (compiled C) and external (written in SC's C-like language) routines. Permutation-based methods for contingency tables (chisquare, likelihood, Kendall S, Theil U, kappa, tau, odds ratio), one and two-sample inference for both means and variances, correlation, and multivariate analysis (MV runs test, Boyett/Schuster, Hoetelling's T. Ready-made bootstrap routines for testing homoscedacity, detecting multimodality, plus general bootstrapping and jack-knifing facilities. Dr. Tony Dusoir, Mole Software, 23 Cable Rd., Whitehead, Co. Antrim BT38 9PZ, N. Ireland.

**StatXact** is for anyone who routinely analyzes categorical or ordered data. The StatXact manual is a textbook in its own right. Can perform any of the techniques outlined in Chapter 6. Versions for Windows or Unix. Also available as an add-on module for both SAS and SPSS. **LogXact**, a companion product, supports inference for logistic regression and provides conditional logistic regression for matched case-control studies or clustered binomial data. Cytel Software Corporation, 675 Massachusetts Avenue, Cambridge, MA 02139. 617-661-2011. www.cytel.com

# Bibliography

Adams DC; Anthony CD. Using randomization techniques to analyse behavioural data. *Animal Behav.* 1996; 51: 733–738.

Agresti A. *Categorical Data Analysis.* New York: John Wiley & Sons; 1990.

Agresti A. A survey of exact inference for contingency tables. *Stat. Sci.* 1992; 7: 131–177.

Albers W; Bickel PJ; & Van Zwet WR. Asymptotic expansions for the power of distribution-free tests in the one-sample problem. *Ann. Statist.* 1976; 4: 108–156.

Alderson MR; Nayak RA. Study of space-time clustering in Hodgkin's disease in the Manchester Region. *Brit. J. Preventive and Social Medicine.* 1971; 25:168–173.

Aly E-E AA. Simple tests for dispersive ordering. *Stat. Prob. Ltr.* 1990; 9: 323–325.

Antretter E; Dunkel D; & Haring C. The WHO/EURO multi-centre study of suicidal behaviour: Findings of the Austrian research centre in Europe-wide comparison. *Wien Klin Wochenschr* 2000; 112: 955–964.

Arndt S; Cizadlo T; Andreasen NC; Heckel D; Gold S; & Oleary DS. Tests for comparing images based on randomization and permutation methods. *J. Cerebral Blood Flow and Metabolism.* 1996; 16: 1271–1279.

Baglivo J; Olivier D; & Pagano M. Methods for the analysis of contingency tables with large and small cell counts. *JASA.* 1988; 83: 1006–1013.

Baker RD. Two permutation tests of equality of variance. *Stat. & Comput.* 1995; 5: 289–296.

Barbe P; Bertail P. *The Weighted Bootstrap.* New York: Springer-Verlag; 1995.

Barbella P; Denby L; & Glandwehr JM. Beyond exploratory data analysis: The randomization test. *Math Teacher.* 1990; 83: 144–149.

218    Bibliography

Barton DE; David FN. Randomization basis for multivariate tests. *Bull. Int. Statist. Inst.* 1961; 39(2): 455–467.

Basu D. Discussion of Joseph Berkson's paper "In dispraise of the exact test." *J. Statist. Plan. Infer.* 1979; 3: 189–192.

Berger JO; Wolpert RW. *The Likelihood Principle.* IMS Lecture Notes—Monograph Series. Hayward CA:IMS; 1984.

* Berger VW. Pros and cons of permutation tests. *Statistics in Medicine* 2000; 19: 1319–1328.

Berkson J. In dispraise of the exact test. *J. Statist. Plan. Inf.* 1978; 2: 27–42.

Berry KJ; Kvamme KL; & Mielke PW Jr. Permutation techniques for the spatial analysis of the distribution of artifacts into classes. *Amer. Antiquity.* 1980; 45: 55–59.

Berry KJ; Kvamme KL; & Mielke PW Jr. Improvements in the permutation test for the spatial analysis of the distribution of artifacts into classes. *Amer. Antiquity.* 1983; 48: 547–553.

Bickel PM; Van Zwet WR. Asymptotic expansion for the power of distribution free tests in the two-sample problem. *Ann. Statist.* 1978; 6: 987–1004 (corr 1170–1171).

Bishop YMM; Fienberg SE; & Holland PW. *Discrete Multivariate Analysis: Theory and Practice.* Cambridge MA: MIT Press; 1975.

Blair C; Higgins JJ; Karinsky W; Krom R; & Rey JD. A study of multivariate permutation tests which may replace Hoetellings T test in prescribed circumstances. *Multivariate Beh. Res.* 1994; 29: 141–163.

Blair RC; Troendle JF; & Beck RW. Control of familywise errors in multiple endpoint assessments via stepwise permutation tests. *Statistics in Medicine.* 1996; 15: 1107–1121.

Boess FG; Balasuvramanian R; Brammer MJ; & Campbell IC. Stimulation of muscarinic acetylcholine receptors increases synaptosomal free calcium concentration by protein kinase-dependent opening of L-type calcium channels. *J. Neurochem.* 1990; 55: 230–236.

Boyett JM; Shuster JJ. Nonparametric one-sided tests in multivariate analysis with medical applications. *JASA.* 1977; 72: 665–668.

Bradbury IS. Analysis of variance vs randomization tests: a comparison (with discussion by White and Still). *Brit. J. Math. Stat. Psych.* 1987; 40: 177–195.

Bradley JV. *Distribution Free Statistical Tests.* New Jersey: Prentice Hall; 1968.

Breiman L. The little bootstrap and other methods for density selection in regression.: X-fixed prediction error. *JASA.* 1992; 87: 738–754.

Breiman L; Friedman JH; Olshen RA; & Stone CJ. *Classification and Regression Trees.* Belmont CA: Wadsworth; 1984.

Bross IDJ. Taking a covariable into account. *JASA.* 1964; 59: 725–736.

Bryant EH. Morphometric adaptation of the housefly, Musa domestica L; in the United States. *Evolution.* 1977; 31: 580–596.

Bullmore E; Brammer M; et al. Statistical methods for estimation and inference for functional MR image analysis. *Magn. Res M.* 1996; 35: 261–727.

Buonaccorsi JP. A note on confidence intervals for proportions in finite populations. *Amer. Stat.* 1987; 41: 215–218.

Burgess AP. Gruzelier JH. Short duration power changes in the EEG during recognition memory for words and faces. *Psychophysiology.* 2000; 37: 596–606.

Busby DG. Effects of aerial spraying of fenithrothion on breeding white-throated sparrows. *J. Appl. Ecol.* 1990; 27: 745–755.

Cade B. Comparison of tree basal area and canopy cover in habitat models: Subalpine forest. *J. Alpine Mgmt.* 1997; 61: 326–335.

Cade B; Hoffman H. Differential migration of blue grouse in Colorado. *Auk.* 1993; 110:70–77.

Cade B; Richards L. Permutation tests for least absolute deviation regression. *Biometrics.* 1996; 52: 886–902.

Carpenter J; Bithell J. Bootstrap confidence intervals: when, which, what? A practical guide for medical statisticians. *Statistics in Medicine.* 2000; 19:1141–1164.

Carter C; Catlett J. Assessing credit card applications using machine learning. *IEEE Expert.* 1987; 2: 71–79.

Chernick MR. *Bootstrap Methods : A Practitioner's Guide.* New York:Wiley; 1999.

Chhikara RK. State of the art in credit evaluation. *Amer. J. Agric. Econ.* 1989; 71: 1138–1144.

Cleveland WS. *The Elements of Graphing Data.* Monterey CA: Wadsworth; 1985.

Cleveland WS. *Visualizing Data.* Summit NJ: Hobart Press; 1993.

Cliff AD; Ord JK. Evaluating the percentage points of a spatial autocorrelation coefficient. *Geog. Anal.* 1971; 3: 51–62.

Cliff AD; Ord JK. *Spatial Processes: Models and Applications.* London: Pion Ltd; 1981.

Cox DR; Snell EJ. *Applied Statistics. Principles and Examples.* London: Chapman and Hall; 1981.

Cover T; Hart P. Nearest neighbor pattern classification. *IEEE Trans. on Info. Theory.* 1967; IT-13: 21–27.

Dalen J. Computing elementary aggregates in the Swedish Consumer Price Index. *J Official Statist.* 1992; 8: 129–147.

Davis AW. On the effects of moderate nonnormality on Roy's largest root test. *JASA.* 1982; 77: 896–900.

DeJager OC; Swanepoel JWH; & Raubenheimer BC. Kernel density estimators applied to gamma ray light curves. *Astron. & Astrophy.* 1986; 170: 187–196.

Diaconis P; Efron B. Computer intensive methods in statistics. *Sci. Amer.* 1983; 48: 116–130.

DiCiccio TJ; Hall P; & Romano JP. On smoothing the bootstrap. *Ann. Statist.* 1989; 17: 692–704.

DiCiccio TJ; Romano J. A review of bootstrap confidence intervals (with discussions). *JRSS B.* 1988; 50: 163–170.

Dietz EJ. Permutation tests for the association between two distance matricies. *Systemic Zoology.* 1983; 32: 21–26.

Diggle PJ; Lange N; & Benes FM. Analysis of variance for replicated spatial point paterns in Clinical Neuroanatomy. *JASA.* 1991; 86: 618–625.

Do KA; Hall P. On importance resampling for the bootstrap. *Biometrika.* 1991; 78: 161–167.

Doolittle RF. Similar amino acid sequences: chance or common ancestory. *Science.* 1981; 214: 149–159.

Douglas ME; Endler JA. Quantitative matrix comparisons in ecological and evolutionary investigations. *J. Theoret. Biol.* 1982; 99: 777–795.

Draper D; Hodges JS; Mallows CL; & Pregibon D. Exchangeability and data analysis (with discussion). *JRSS A.* 1993; 156: 9–28.

Dubuisson B; Lavison P. Surveillance of a nuclear reactor by use of a pattern recognition methodology. *IEEE Trans. Systems, Man, & Cybernetics.* 1980; 10: 603–609.

Edgington ES. *Randomization Tests.* 3rd ed. New York: Marcel Dekker; 1995.

Efron B. Bootstrap methods: Another look at the jackknife. *Annals Statist.* 1979; 7: 1–26.

Efron B. *The Jackknife, the Bootstrap and Other Resampling Plans.* Philadelphia: SIAM; 1982.

Efron B. Estimating the error rate of a prediction rule: improvements on cross-validation. *JASA.* 1983; 78: 316–331.

Efron B. Bootstrap confidence intervals: good or bad? (with discussion). *Psychol Bull.* 1988; 104: 293–296.

Efron B. Six questions raised by the bootstrap. R. LePage and L. Billard, eds. *Exploring the Limits of the Bootstrap.* New York: Wiley; 1992.

Efron B; DiCiccio T. More accurate confidence intervals in exponential families. *Biometrika.* 1992; 79: 231–245.

Efron B; Tibshirani R. Bootstrap measures for standard errors, confidence intervals, and other measures of statistical accuracy. *Stat Sci.* 1986; 1: 54–77.

Efron B; Tibshriani R. Statistical data analysis in the computer age. *Science.* 1991; 253: 390–395.

Efron B; Tibshirani R. *An Introduction to the Bootstrap*. New York: Chapman and Hall; 1993.

Entsuah AR. Randomization procedures for analyzing clinical trend data with treatment related withdrawls. *Comm. Statist. A*. 1990; 19: 3859–3880.

Falk M; Reiss RD. Weak convergence of smoothed and nonsmoothed bootstrap quantiles estimates. *Ann. Prob*. 1989; 17: 362–371.

Faris PD; Sainsbury RS. The role of the Pontis Oralis in the generation of RSA activity in the hippocampus of the guinea pig. *Psych.& Beh*. 1990; 47: 1193–1199.

Farrar DA; Crump KS. Exact statistical tests for any coarcinogenic effect in animal assays. *Fund. Appl. Toxicol*. 1988; 11: 652–663.

Farrar DA; Crump KS. Exact statistical tests for any coarcinogenic effect in animal assays. II age adjusted tests. *Fund. Appl. Toxicol*. 1991; 15: 710–721.

Fears TR; Tarone RE; & Chu KC. False-positive and false-negative rates for carcinogenicity screens. *Cancer Res*. 1977; 37: 1941–1945.

Feinstein AR. Clinical biostatistics XXIII. The role of randomization in sampling, testing, allocation, and credulous idolatry (part 2). *Clinical Pharm*. 1973; 14: 989–1019.

Festinger LC; Carlsmith JM. Cognitive consequences of forced compliance. *J. Abnorm. Soc. Psych*. 1959; 58: 203–210.

Fisher RA. *Statistical Methods for Research Workers*. Edinburgh: Oliver & Boyd; 1st ed 1925.

Fisher RA. The logic of inductive inference (with discussion). *JRSS A*. 1934; 98: 39–54.

Fisher RA. *Design of Experiments*. New York: Hafner; 1935.

Ford RD; Colom LV; & Bland BH. The classification of medial septum-diagonal band cells as theta-on or theta-off in relation to hippo campal EEG states. *Brain Res*. 1989; 493: 269–282.

Forsythe AB; Engleman L; Jennrich R. A stopping rule for variable selection in multivariate regression. *JASA*. 1973; 68: 75–77.

Foutz RN; Jensen DR; & Anderson GW. Multiple comparisons in the randomization analysis of designed experiments with growth curve responses. *Biometrics*. 1985; 41: 29–37.

Frank D; Trzos RJ; & Good P. Evaluating drug-induced chromosome alterations. *Mutation Res*. 1978; 56: 311–317.

Fraumeni JF; Li FP. Hodgkin's disease in childhood: an epidemiological study. *J. Nat. Cancer Inst*. 1969; 42: 681–691.

Freedman DA. A note on screening regression equations. *Amer. Statist*. 1983; 37: 152–155.

Freedman D; Diaconis P. On the histogram as a density estimator: L2 theory. *Zeitschrift fur Wahrscheinlichkeitstheorie und verwandte Gebeite.* 1981; 57: 453–476.

Freeman GH; Halton JH. Note on an exact treatment of contingency, goodness of fit, and other problems of significance. *Biometrika.* 1951; 38: 141–149.

Gabriel KR. Some statistical issues in weather experimentation. *Commun Statist. A.* 1979; 8: 975–1015.

Gabriel KR; Hsu CF. Evaluation of the power of rerandomization tests, with application to weather modification experiments. *JASA.* 1983; 78: 766–775.

Gabriel KR; Sokal RR. A new statistical approach to geographical variation analysis. *Systematic Zoology.* 1969; 18: 259–270.

Gail M; Mantel N. Counting the number of rxc contingency tables with fixed marginals. *JASA.* 1977; 72: 859–862.

Gail MH; Tan WY; & Piantadosi S. Tests for no treatment effect in randomized clinical trials. *Biometrika.* 1988; 75: 57–64.

Gallant AR. *Nonlinear Statistical Models.* New York: Wiley. 1987.

Gart JJ. *Statistical Methods in Cancer Res., Vol III - The design and analysis of long term animal experiments.* Lyon: IARC Scientific Publications; 1986.

Garthwaite PH. Confidence intervals from randomization tests. *Biometrics.* 1996; 52: 1387–1393.

Gastwirht JL. Statistical reasoning in the legal setting. *Amer. Statist.* 1992; 46: 55–69.

Geisser S. The predictive sample reuse method with applications. *JASA.* 1975; 70: 320–328.

Gine E; Zinn J. Necessary conditions for a bootstrap of the mean. *Ann. Statist.* 1989; 17:684–691.

Glass AG; Mantel N. Lack of time-space clustering of childhood leukemia, Los Angeles County 1960–64. *Cancer Res.* 1969; 29: 1995–2001.

Glass AG; Mantel N; Gunz FW; & Spears GFS. Time-space clustering of childhood leukemia in New Zealand. *J. Nat. Cancer Inst.* 1971; 47: 329–336.

Gleason JR. Algorithms for balanced bootstrap simulations. *Amer. Statist.* 1988; 42: 263–266.

Gliddentracey C; Greenwood AK. A validation study of the Spanish self directed search using back translation procedures. *J. Career Assess.* 1997; 5: 105–113.

Gliddentracey CE; Parraga MI. Assessing the structure of vocational interests among Bolivian university students. *J. Vocational Beh.* 1996; 48: 96–106.

Gonser R; Donnelly P; Nicholson G. et al. Microsatellite mutations and inferences about human demography. *Genetics* 2000; 154: 1793–1807.

Good PI. Detection of a treatment effect when not all experimental subjects respond to treatment. *Biometrics.* 1979; 35: 483–489.

Good PI. Almost most powerful tests for composite alternatives. *Comm. Statist.—Theory & Methods. 1*989; 18(5): 1913–1925.

Good PI. Most powerful tests for use in matched pair experiments when data may be censored. *J. Statist. Comp. Simul.* 1991; 38: 57–63.

Good PI. Globally almost powerful tests for censored data. *Nonpar. Statist.* 1992; 1: 253–262.

‛ Good PI. *Permutation Tests.* New York: Springer Verlag; 2nd ed, 2000.

⟋ Good PI. *Applying Statistics in the Courtroom.* London: Chapman & Hall; 2001.

Goodman L; Kruskal W. Measures of association for cross-classification. *JASA.* 1954; 49: 732–764.

Graubard BI; Korn EL. Choice of column scores for testing independence in ordered 2 by K contingency tables. *Biometrics.* 1987; 43: 471–476.

Graves GW; Whinston AB. An algorithm for the quadratic assignment probability. *Mgmt. Science.* 1970; 17: 453–471.

Grossman DC; Cummings P; Koepsell TD. et al. Firearm safety counseling in primary care pediatrics: A randomized. controlled trial. *Pediatrics* 2000; 106: 22–26.

Gupta et al. *Community Dentistry and Oral Epidemiology.* 1980; 8: 287–333.

Haber M. A comparison of some conditional and unconditional exact tests for 2x2 contingency tables. *Comm. Statist. A.* 1987; 18: 147–156.

Hall P. On the bootstrap and confidence intervals. *Ann. Statist.* 1986; 14: 1431–1452.

Hall P. Theoretical comparison of bootstrap confidence intervals (with discussion). *Ann. Statist.* 1988; 16: 927–985.

Hall P; Hart JD. Bootstrap test for difference between means in nonparametric regression. *JASA.* 1990; 85:1039–1049.

Hall P; Martin MA. On bootstrap resampling and iteration. *Biometrika.* 1988; 75: 661–671.

Hall P; Titterington M. The effect of simulation order on level accuracy and power of Monte Carlo tests. *JRSS B.* 1989; 51: 459–467.

Hall P; Wilson SR. Two guidelines for bootstrap hypothesis testing. *Biometrics.* 1991; 47: 757–762.

Halter JH. A rigorus derivation of the exact contingency formula. *Proc. Cambridge Phil. Soc.* 1969; 65: 527–530.

Hårdle W. *Smoothing Techniques with Implementation in S.* New York: Springer-Verlag; 1991.

Hardy FL;Youse BK. *Finite Mathematics for the Managerial, Social, and Life Sciences.* New York: West Publishing; 1984.

Hartigan JA. Using subsample values as typical values. *JASA.* 1969; 64: 1303–1317.

Hartigan JA. Error analysis by replaced samples. *J. Roy. Statist. Soc B.* 1971; 33: 98–110.

Hasegawa M; Kishino H; & Yano T. Phylogenetic inference from DNA sequence data. K. Matusita, editor. *Statistical Theory and Data Analysis.* Amsterdam: North Holland; 1988.

Henze N. A multivariate two-sample test based on the number of nearest neighbor coincidence. *Annals Statist.* 1988; 16: 772–783.

Hettmansperger TP. *Statistical Inference Based on Ranks.* New York: Wiley; 1984.

Highton R. Comparison of microgeographic variation in morphological and electrophoretic traits. In Hecht MK, Steer WC, & B Wallace eds. *Evolutionary Biology* New York: Plenum; 1977; 10: 397–436.

Hill AB. *Principles of Medical Statistics* (8th ed). London: Oxford University Press; 1966.

Hinkley DV; Shi S. Importance sampling and the nested bootstrap. *Biometrika.* 1989; 76: 435–446.

Hirji KF; Mehta CR; & Patel NR. Computing distributions for exact logistic regression. *JASA.* 1987; 82: 1110–1117.

Hisdal H; Stahl K; Tallaksen LM. et al. Have streamflow droughts in Europe become more severe or frequent? *Int J Climatol.* 2001; 21 : 317–321.

Hjorth JSU. *Computer Intensive Statistical Methods: Validation, Model Selection and Bootstrap.* New York: Chapman and Hall; 1994.

Hollander M; Pena E. Nonparametric tests under restricted treatment assignment rules. *JASA.* 1988; 83: 1144–1151.

Hollander M; Sethuraman J. Testing for agreement between two groups of judges. *Biometrika.* 1978; 65: 403–412.

Holmes MC; Williams REO. The distribution of carriers of streptococcus pyrogenes among 2413 healthy children. *J. Hyg. Camd.* 1954; 52: 165–179.

Hosmer DW; Lemeshow S. Best subsets logistic regression. *Biometrics.* 1989.

Howard M (pseud for Good P). Randomization in the analysis of experiments and clinical trials. *American Laboratory.* 1981; 13: 98–102.

Hubert LJ. Combinatorial data analysis: Association and partial association. *Psychometrika.* 1985; 50: 449–467.

Hubert LJ; Baker FB. Analyzing distinctive features confusion matrix. J. Educ. Statist. 1977; 2: 79–98.

Hubert LJ; Baker FB. Evaluating the conformity of sociometric measurements. *Psychometrika.* 1978; 43: 31–42.

Hubert LJ; Golledge RG; & Costanzo CM. Analysis of variance procedures based on a proximity measure between subjects. *Psych. Bull.* 1982; 91: 424–30.

Hubert LJ; Golledge RG; Costanzo CM; Gale N; & Halperin WC. Nonparametric tests for directional data. In Bahrenberg G, Fischer M, &. P. Nijkamp, eds. *Recent Developments In Spatial Analysis: Methodology , Measurement, Models.* Aldershot UK: Gower; 1984: 171–190.

Hubert LJ; Schultz J. Quadratic assignment as a general data analysis strategy. *Brit. J. Math. Stat. Psych.*1976; 29: 190–241.

Ingenbleek JF. Tests simultanes de permutation des rangs pour bruit-blanc multivarie. *Statist. Anal Donnees.* 1981; 6: 60–65.

Jackson DA. Ratios in acquatic sciences: Statistical shortcomings with mean depth and the morphoedaphic index. *Canadian J Fisheries and Acquatic Sciences.* 1990; 47: 1788–1795.

Jin MZ. On the multisample pemutation test when the experimental units are nonuniform and random experimental errors exist. *J System Sci Math Sci.* 1984; 4: 117–127, 236–243.

John RD; Robinson J. Significance levels and confidence intervals for randomization tests. *J. Statist. Comput. Simul.* 1983; 16: 161–173.

Johns MV Jr. Importance sampling for bootstrap confidence intervals. *JASA.* 1988; 83: 709–714.

Jones HL. Investigating the properties of a sample mean by employing random subsample means. *JASA.* 1956; 51: 54–83.

Jorde LB; Rogers AR; Bamshad M; Watkins WS; Krakowiak P; Sung S; Kere J; & Harpending HC. Microsatellite diversity and the demographic history of modern humans. *Proc. Nat. Acad. Sci.* 1997; 94: 3100–3103.

Karlin S; Ghandour G; Ost F; Tauare S; & Korph K. New approaches for computer analysis of DNA sequences. *Proc. Nat. Acad. Sci.*, USA. 1983; 80: 5660–5664.

Karlin S; Williams PT. Permutation methods for the structured exploratory data analysis (SEDA) of familial trait values. *Amer. J. Human Genetics.* 1984; 36: 873–898.

Kazdin AE. Statistical analysis for single-case experimental designs. In *Strategies for Studying Behavioral Change.* M Hersen and DH Barlow, eds. New York: Pergammon; 1976.

Kazdin AE. Obstacles in using randomization tests in single-case experiments. *J Educ Statist.* 1980; 5: 253–260.

Keller-McNulty S.; Higgens JJ. Effect of tail weight and outliers on power and type I error of robust permuatation tests for location. *Commun. Stat.—Theory and Methods.* 1987; 16: 17–35.

Kempthorne O. The randomization theory of experimental inference. *JASA.* 1955; 50: 946–967.

Kempthorne O. Some aspects of experimental inference. *JASA.* 1966; 61: 11–34.

Kempthorne O. Inference from experiments and randomization. In *A Survey of Statistical Design and Linear Models.* JN Srivastava, ed. Amsterdam: North Holland; 1975: 303–332.

• Kempthorne O. Why randomize? *J. Statist. Prob. Infer.* 1977; 1: 1–26.

Kempthorne O; Doerfler TE. The behavior of some significance tests under experimental randomization. *Biometrika.* 1969; 56: 231–248.

Klauber MR. Two-sample randomization tests for space-time clustering. *Biometrics.* 1971; 27: 129–142.

Klauber MR; Mustacchi A. Space-time clustering of childhood leukemia in San Francisco. *Cancer Res.* 1970; 30: 1969–1973.

Knight K. On the bootstrap of the sample mean in the infinite variance case. Annal Statist. 1989; 17: 1168–1173.

Koch G (ed). *Exchangeability in Probability and Statistics.* Amsterdam: North Holland; 1982.

Koziol J; Gail M; & Green SB. Testing for interaction in an IxJxK contingency table. *Biometrika.* 1977; 64:271–275.

Krewski D; Brennan J; & M Bickis. The power of the Fisher permutation test in 2 by k tables. *Commun. Stat. B.* 1984; 13: 433–448.

Kryscio RJ; Meyers MH; Prusiner SI; Heise HW; & Christine BW. The space-time distribution of Hodgkin's disease in Connecticut, 1940–1969. *J. Nat. Cancer Inst.* 1973; 50: 1107–1110.

Lachin JM. Properties of sample randomization in clinical trials. *Contr. Clin. Trials.* 1988; 9: 312–326.

Lachin JN. Statistical properties of randomization in clinical trials. *Contr. Clin. Trials.* 1988; 9: 289–311.

Lahri SN. Bootstrapping the studentized sample mean of lattice variables. *J. Multiv. Anal.* 1993; 45:247–256.

Laitenberger O; Atkinson C; Schlich M. et al. An experimental comparison of reading techniques for defect detection in UML design documents. *J Syst Software.* 2000; 53: 183–204.

Lambert D. Robust two-sample permutation tests. *Ann. Statist.* 1985; 13: 606–625.

Leemis LM. Relationships among common univariate distributions. *Amer. Statist.* 1986; 40: 143–146.

Lefebvre M. Une application des methodes sequentielles aux tests de permutations. *Canad. J. Statist.* 1982; 10: 173–180.

Lehmann EL. *Testing Statistical Hypotheses.* New York: John Wiley & Sons; 1986.

Levin DA. The organization of genetic variability in Phlox drummondi. *Evolution.* 1977; 31: 477–494.

Liu RY. Bootstrap procedures under some non iid models. *Ann. Statist.* 1988; 16: 1696–1788.

Loughin TM; Noble W. A permutation test for effects in an unreplicated factorial design. *Technometrics.* 1997; 39: 180–190.

Lunneborg CE. Estimating the correlation coefficient: The bootstrap approach. *Psychol. Bull.* 1985; 98: 209–215.

Makinodan T; Albright JW; Peter CP; Good PI; & Hedrick ML. Reduced humoral activity in long-lived mice. *Immunology.* 1976; 31: 400–408.

Makutch RW; Parks WP. Response to serum antigen level to AZT for the treatment of AIDS. *AIDS Research and Human Retroviruses.* 1988; 4: 305–316.

Manly BFJ. *Randomization, Bootstrap and Monte Carlo Methods in Biology.* (2nd ed.). London: Chapman & Hall; 1997.

Mantel N. The detection of disease clustering and a generalized regression approach. *Cancer Res.* 1967; 27: 209–220.

Mantel N; Bailar JC. A class of permutational and multinomial tests arising in epidemiological research. *Biometrics.* 1970; 26: 687–700.

Mapleson WW. The use of GLIM and the bootstrap in assessing a clinical trial of two drugs. *Statist. Med.* 1986; 5: 363–374.

Marcus LF. Measurement of selection using distance statistics in prehistoric orangutan pongo pygamous palaeosumativens. *Evolution.* 1969; 23: 301.

Mardia KV; Kent JT; Bibby JM. *Multivariate Analysis.* New York: Academic Press; 1979.

Maritz JS. *Distribution Free Statistical Methods.* (2nd ed.) London: Chapman & Hall; 1996.

Marron JS. A comparison of cross-validation techniques in density estimation. *Ann. Statist.* 1987; 15: 152–162.

Martin MA. On bootstrap iteration for coverage of confidence intervals. *JASA.* 1990; 85:1105–1108.

Martin-Lof P. Exact tests, confidence regions and estimates. In Barndorff-Nielsen O; Blasild P; & Schow G., eds. *Proceeding of the Conference of Foundational Questions in Statistical Inference.* Aarhus: Institute of Mathematics, University of Aarhus; 1974; 1: 121–138.

Maxwell SE; Cole DA. A comparison of methods for increasing power in randomized between-subjects designs. *Psych Bull.* 1991; 110: 328–337.

McCarthy PJ. Psuedo-replication: Half samples. *Review Int. Statist. Inst.* 1969; 37: 239–264.

McKinney PW; Young MJ; Hartz A; Bi-Fong Lee M. The inexact use of Fisher's exact test in six major medical journals. *J. American Medical Association.* 1989; 261: 3430–3433.

McLachlan G. *Finite Mixture Models.* New York: Wiley; 2000.

Mehta CR; Patel NR. A network algorithm for the exact treatment of the 2xK contingency table. *Commun. Statist.* B. 1980; 9: 649–664.

Mehta CR; Patel NR. A network algorithm for performing Fisher's exact test in rxc contingency tables. *JASA.* 1983; 78: 427–434.

Mehta CR; Patel NR; Gray R. On computing an exact confidence interval for the common odds ratio in several 2x2 contingency tables. *JASA.*. 1985; 80: 969–973.

Mehta CR; Patel NR; Senchaudhuri P. Importance sampling for estimating exact probabilities in permutational inference. *JASA.* 1988; 83: 999–1005.

Merrington M; Spicer CC. Acute leukemia in New England. *Brit J Preventive and Social Medicine.* 1969; 23: 124–127.

Micceri T. The unicorn, the normal curve, and other improbable creatures. *Psychol. Bull.* 1989; 105: 156–166.

Mielke PW. Some parametric, nonparametric and permutation inference procedures resulting from weather modification experiments. *Commun Statist. A.* 1979; 8: 1083–1096.

Mielke PW. Meterological applications of permutation techniques based on distance functions. In Krishnaiah PR; Sen PK, eds. *Handbook of Statistics.* Amsterdam: North-Holland; 1984; 4: 813–830.

Mielke PW. Non-metric statistical analysis: Some metric alternatives. *J. Statist. Plan. Infer.* 1986; 13: 377–387.

Mielke PW Jr. The application of multivariate permutation methods based on distance functions in the earth sciences. *Earth-Science Rev.* 1991; 31: 55–71.

Mielke PW Jr; Berry KJ. Fisher's exact probability test for cross-classification tables. *Educational and Psychological Measurement.* 1992; 52: 97–101.

Milano F; Maggi E; del Turco MR. Evaluation of the effect of a quality control programme in mammography on technical and exposure parameters. *Radiat Prot Dosim.* 2000; 90: 263–266.

Mooney CZ; Duval RD. *Bootstrapping: A Nonparametric Approach to Statistical Inference.* Newbury Park CA: Sage Publications; 1993.

Mosteller F; Tukey JW. *Data Analysis and Regression: A second course in statistics.* Menlo Park: Addison-Wesley; 1977.

Mueller LD; Altenberg L. Statistical inference on measures of niche overlap. *Ecology.* 1985; 66: 1204–1210.

Neyman J. *First Course in Probability and Statistics.* New York: Holt, 1950.

Nguyen TT. A generalization of Fisher's exact test in pxq contingency tables using more concordant relations. *Commun. Statist. B.* 1985; 14: 633–645.

Noether GE. Distribution-free confidence intervals. *Statistica Neerlandica.* 1978; 32: 104–122.

Noreen E. *Computer Intensive Methods for Testing Hypotheses.* New York: Wiley, 1989.

Pagan A; Ullah A. *Nonparametric Economics.* Cambridge: Cambridge University Press, 1999.

Oden A; Wedel H. Arguments for Fisher's permutation test. *Ann. Statist.* 1975; 3: 518–520.

Passing H. Exact simultaneous comparisons with controls in an rxc contingency table. *Biometrical J.* 1984; 26: 643–654.

Patefield WM. Exact tests for trends in ordered contingency tables. *Appl. Statist.* 1982; 31: 32–43.

Patil CHK. Cochran's Q test: exact distribution. *JASA.* 1975; 70: 186–189.

Pearson ES. Some aspects of the problem of randomization. *Biometrika.* 1937; 29: 53–64.

Penninckx W; Hartmann C; Massart DL; & Smeyersverbeke J. Validation of the calibration procedure in atomic absorption spectrometric methods. *J Analytical Atomic Spectrometry.* 1996; 11: 237–246.

Peritz E. Exact tests for matched pairs: studies with covariates. *Commun. Statist. A.* 1982; 11: 2157–2167 (errata 12: 1209–1210).

Peritz E. Modified Mantel-Haenszel procedures for matched pairs. *Commun. Statist. A.* 1985; 14: 2263–2285.

Pesarin F. On a nonparametric combination method for dependent permutation tests with applications. *Psychotherapy and Psychosomatics.* 1990; 54:172–179.

Pesarin F. A nonparametric combination method for dependent permutation tests with application to some problems with repeated measures. In *Industrial Statistics,* Kitsos CP and Edler L, eds. Physics-Verlag: Heidelberg, 1997; 259–268.

Pesarin F. *Multivariate Permutation Tests.* NewYork: Wiley, 2001.

Petrondas DA; Gabriel RK. Multiple comparisons by rerandomization tests. *JASA.* 1983; 78: 949–957.

Pitman EJG. Significance tests which may be applied to samples from any population. *Roy. Statist. Soc. Suppl.* 1937; 4: 119–130, 225–232.

Pitman EJG. Significance tests which may be applied to samples from any population. Part III. The analysis of variance test. *Biometrika.* 1938; 29: 322–335.

Plackett RL; Hewlett PS. A unified theory of quantal responses to mixtures of drugs. The fitting to data of certain models for two non-interactive drugs with complete positive correlation of tolerances. *Biometrics.* 1963; 19: 517–531.

Pollard E; Lackland KH; & Rothrey P. The detection of density dependence from a series of annual censuses. *Ecology.* 1987; 68: 2046–2055.

Prager MH; Hoenig JM. Superposed epoch analysis: A randomization test of environmental effects on recruitment with application to chub mackrel. *Trans. Amer. Fisheries Soc.* 1989; 18: 608–619.

Praska Rao BLS. *Nonparametric Functional Estimation.* New York: Academic Press, 1983.

Priesendorfer RW; Barnett TP. Numerical model/reality intercomparison tests using small-sample statistics. *J. Atmospheric Sci.* 1983; 40: 1884–96.

Puri ML; Sen PK. On a class of multivariate, multisample rank-order tests. *Sankyha Series A.* 1966; 28: 353–376.

Ratkowsky DA. *Handbook of Nonlinear Regression Models*. New York: Dekker, 1990.

Rasmussen J. Estimating correlation coefficients: bootstrap and parametric approaches. *Psych. Bull.* 1987; 101: 136–139.

Raz J. Tesῠing for no effect when estimating a smooth function by nonparametric regression: a randomization approach. *JASA*. 1990; 85: 132–138.

Romano JP. A bootstrap revival of some nonparametric distance tests. *JASA*. 1988; 83: 698–708.

Romano JP. On the behavior of randomization tests without a group invariance assumption. *JASA.*. 1990; 85(411): 686–692.

Rosenbaum PR. Permutation tests for matched pairs with adjustments for covariates. *Appl. Statist.* 1988; 37: 401–411.

Roy J. *Ann. Math. Statist.* 1957; 29: 1177–1187.

Royaltey HH; Astrachen E; Sokal RR. Tests for patterns in geographic variation. *Geographic Analysis*. 1975; 7: 369–395.

Runger GC; Eaton MI. Most powerful invariant tests. *J. Multiv. Anal.* 1992; 42: 202–209.

Ryan JM; Tracey TJG; & Rounds J. Generalizability of Holland's structure of vocational interests across ethnicity, gender, and socioeconomic status. *J. Counseling Psych.* 1996; 43: 330–337.

Ryan TP. *Modern Regression Methods*. New York: Wiley, 1997.

Ryman N; Reuterwall C; Nygren K; & Nygren T. Genetic variation and differentiation in Scandiavian moose (Alces Alces): Are large mammals monomorphic? *Evolution*. 1980; 34: 1037–1049.

Salapatek P; Kessen W. Visual scanning of triangles by the human newborn. *J Exper Child Psych.* 1966; 3: 155–167.

Scheffe H. *Analysis of Variance*. New York: Wiley Sons, 1959.

Schenker N. Qualms about bootstrap confidence intervals. *JASA* 1985; 80: 360–361.

Schimek MG. *Smoothing and Regression*. New York: Wiley, 2000.

Schultz JR; Hubert L. A nonparametric test for the correspondence between two proximity matrices. *J. Educ. Statist.* 1976; 1: 59–67.

Scott DW. *Multivariate Density Estimation: Theory, Practice, and Visualization*. New York: Wiley, 1992.

Scott DW; Gotto AM; Cole JS: & Gorry JA. Plasma lipids as collateral risk factors in coronary artery disease—a study of 371 males with chest pain. *J. Chronic Diseases*. 1978; 31: 337–345.

Selander RK; Kaufman DW. Genetic structure of populations of the brown snail (Helix aspersa). I Microgeographic variation. *Evolution*. 1975; 29: 385–401.

Shao J; Tu D. *The Jacknife and the Bootstrap*. New York: Springer, 1995.

Shen CD; Quade D. A randomization test for a three-period three-treatment cross-over experiment. *Commun. Statist*. B. 1986; 12: 183–199.

Shimabukuro FI; Lazar S; Dyson HB; & Chernick MR. A quasi-optical method for measuring the complex permittivity of materials. *IEEE Transactions on Microwave Theory and Technology*. 1984; 32: 659–665.

Shuster JJ. *Practical Handbook of Sample Size Guidelines for Clinical Trials*. Boca Raton FL: CRC Press, 1993.

Shuster JJ; Boyett JM. Nonparametric multiple comparison procedures. *JASA*.. 1979; 74: 379–382.

Siemiatycki J. Mantel's space-time clustering statistic: computing higher moments and a comparison of various data transforms. *J Statist. Comput. Simul*. 1978; 7: 13–31.

Siemiatycki J; McDonald AD. Neural tube defects in Quebec: A search for evidence of 'clustering' in time and space. *Brit. J. Prev. Soc. Med*. 1972; 26: 10–14.

Silverman BW. Using kernel density estimates to investigate multimodality. JRSS B. 1981; 43: 97–99.

Silverman BW; Young GA. The bootstrap: to smooth or not to smooth. *Biometrika*. 1987; 74: 469–479.

Simon JL. *Basic Research Methods in Social Science*. New York: Random House; 1969.

Singh K. On the asymptotic accuracy of Efrons bootstrap. *Ann. Statist*. 1981; 9: 1187–1195.

Smith PG; Pike MC. Generalization of two tests for the detection of household aggregation of disease. *Biometrics*. 1976; 32: 817–828.

Smith PWF; Forster JJ; & McDonald JW. Monte Carlo exact tests for square contingency tables. *J. Royal Statist. Soc. A*. 1996; 159: 309–21.

Smythe RT. Conditional inference for restricted randomization designs. *Ann. Math. Statist*. 1988; 16: 1155–1161.

Sokal RR. Testing statistical significance in geographical variation patterns. *Systematic Zoo*. 1979; 28: 227–232.

Solomon H. Confidence intervals in legal settings. In *Statistics and The Law*. DeGroot MH; Fienberg SE; & Kadane JB, ed. New York: Wiley, 1986: 455–473.

Solow AR. A randomization test for misclassification problems in discriminatory analysis. *Ecology*. 1990; 71: 2379–2382.

Soms AP. Permutation tests for k-sample binomial data with comparisons of exact and approximate P-levels. *Commun. Statist. A*. 1985; 14: 217–233.

Stilson DW. *Psychology and Statistics in Psychological Research and Theory*. San Francisco: Holden Day, 1966.

Stine R. An introduction to bootstrap methods: examples and ideas. In *Modern Methods of Data Analysis*. J. Fox & JS Long, eds. Newbury Park CA: Sage Publications; 1990: 353–373.

Stockmarr A. The choice of hypotheses in the evaluation of DNA profile evidence. In *Statistical Science in the Courtroom*. JL Gastwirth, ed. New York: Springer; 2000.

Stone M. Cross-validation choice and assessment of statistical predictions. *JASA*. 1974; B36: 111–147.

"Student." Errors of routine analysis. *Biometrika*. 1927; 19: 151–164.

Suissa S; Shuster JJ. Are uniformly most powerful unbiased tests really best? *Amer. Statistician*. 1984; 38: 204–206.

Syrjala SE. A statistical test for a difference between the spatial distributions of two populations. *Ecology*. 1996; 77, 75–80.

Thaler-Neto A; Fries R; & Thaller G. Risk ratio as parameter for the genetic characterization of complex binary traits in cattle. A simulation study under various genetic models using halfsib families. *J Anim Breed. Genetics*. 2000; 117: 153–167.

Thompson JR; Bartoszynski R; Brown B; & McBride C. Some nonparametric techniques for estimating the intensity function of a cancer related nonstationary Poisson process. *Ann. Statist*. 1981; 9: 1050–1060.

Thompson JR.; Bridges E; & Ensor K. Marketplace competition in the personal computer industry. *Decision Sciences*. 1992: 467–477.

Thompson JR.; Tapia RA. *Nonparametric Function Estimation, Modeling and Simulation*. Philadelphia PA: SIAM, 1990.

Tibshirani RJ. Variance stabilization and the bootstrap. *Biometrika*. 1988; 75: 433–444.

Titterington DM; Murray GD; Spiegelhalter DJ; Skene AM; Habbema JDF; & Gelke GJ. Comparison of discrimination techniques applied to a complex data set of head-injured patients. JRSS A. 1981; 144: 145–175.

Tracy DS; Khan KA. Comparison of some MRPP and standard rank tests for three equal sized samples. *Commun. Statist*. B. 1990; 19: 315–333.

Tracy DS; Tajuddin IH. Empirical power comparisons of two MRPP rank tests. *Commun. Statist*. A. 1986; 15: 551–570.

Tritchler D. On inverting permutation tests. *JASA*. 1984; 79: 200–207.

Troendle JF. A stepwise resampling method of multiple hypothesis testing. *JASA*. 1995; 90: 370–378.

Tsuji R. The structuring of trust relations in groups and the transition of within-group order. *Sociol Theor Method*. 2000; 15: 197–208.

Tsutakawa RK; Yang SL. Permutation tests applied to antibiotic drug resistance. *JASA*. 1974; 69: 87–92.

Tukey JW. Improving crucial randomized experiments—especially in weather modification—by double randomization and rank combination. In LeCam L; & Binckly P, eds. *Proceeding of the Berkeley Conference in Honor of J Neyman and J Kiefer.* Hayward CA: Wadsworth; 1985; 1: 79–108.

Valdes-Perez RE. Some recent human-computer studies in science and what accounts for them. *AI Magazine.* 1995; 16: 37–44.

Valdes-Perez RE; Pericliev V. Computer enumeration of significant implicational universals of kinship terminology. *Cross-Cult Res* 1999; 33: 162–174.

Van der Voet H. Comparing the predictive accuracy of models using a simple randomization test. *Chemometrics and Intelligent Laboratory Systems.* 1994; 25: 313–323.

Vankeerberghen P; Vandenbosch C; Smeyers-Verbeke J; & Massart DL. Some robust statistical procedures applied to the analysis of chemical data. *Chemometrics and Intelligent Laboratory Systems.* 1991; 12: 3–13.

Van-Putten B. On the construction of multivariate permutation tests in the multivariate two-sample case. *Statist. Neerlandica.* 1987; 41: 191–201.

Vecchia DF; HK Iyer. Exact distribution-free tests for equality of several linear models. *Commun Statist. A.* 1989; 18: 2467–2488.

Wald A; Wolfowitz J. Statistical tests based on permutations of the observations. *Ann. Math. Statist.* 1944; 15: 358–372.

Wåhrendorf J; Brown CC. Bootstrapping a basic inequality in the analysis of the joint action of two drugs. *Biometrics.* 1980; 36: 653–657.

Wei LJ. Exact two-sample permutation tests based on the randomized play-the-winner rule. *Biometrika.* 1988; 75: 603–605.

Wei LJ; Smythe RT; & Smith RL. K-treatment comparisons in clinical trials. *Annals Math Statist.* 1986; 14: 265–274.

Welch WJ. Construction of permutation tests. *JASA.* 1990; 85(411): 693–698.

Welch WJ. Rerandomizing the median in matched-pairs designs. *Biometrika.* 1987; 74: 609–614.

Welch WJ; Guitierrez LG. Robust permutation tests for matched pairs designs. *JASA.* 1988; 83: 450–461.

Werner M; Tolls R; Hultin J; & Mellecker J. Sex and age dependence of serum calcium, inorganic phosphorous, total protein, and albumin in a large ambulatory population. In *Fifth Technical International Congress of Automation, Advances in Automated Analysis.* Mount Kisco NY: Futura Publishing, 1970.

Westfall DH; Young SS. *Resampling-Based Multiple Testing: Examples and Methods for p-value Adjustment.* New York: John Wiley, 1993.

Whaley FS. The equivalence of three individually derived permutation procedures for testing the homogenity of multidimensional samples. *Biometrics.* 1983; 39: 741–745.

Whittaker J. *Graphical Models in Applied Statistics.* Chichester: Wiley, 1990.

Wilk MB. The randomization analysis of a generalized randomized block design. *Biometrika*. 1955; 42: 70–79.

Wilk MB; Kempthorne O. Nonadditivities in a Latin square design. *JASA*. 1957; 52: 218–236.

Wilk MB; Kempthorne O. Some aspects of the analysis of factorial experiments in a completely randomized design. *Ann. Math. Statist*. 1956; 27: 950–984.

Williams-Blangero S. Clan-structured migration and phenotypic differentiation in the Jirels of Nepal. *Human Biology*. 1989; 61: 143–157.

Witztum D; Rips E; & Rosenberg Y. Equidistant letter sequences in the Book of Genesis. *Statist. Science*. 1994; 89: 768–776.

Wong RKW; Chidambaram N; & Mielke PW. Applications of multi-response permutation procedures and median regression for covariate analyses of possible weather modification effects on hail responses. *Atmosphere-Ocean*. *1983*; 21: 1–13.

Young GA. A note on bootstrapping the correlation coefficient. *Biometrika*. 1988; 75: 370–373.

Young GA. Bootstrap: More than a stab in the dark. *Statist. Science*. 1994; 9: 382–415.

Zempo N; Kayama N; Kenagy RD; Lea HJ; & Clowes AW. Regulation of vascular smooth-muscle-cell migration and proliferation in vitro and in injured rat arteries by a synthetic matrix metalloprotinase inhibitor. *Art Throm V*. 1996; 16: 28–33.

Zerbe GO. Randomization analysis of the completely randomized design extended to growth and response curves. *JASA*. 1979; 74: 215–221.

Zerbe GO. Randomization analysis of randomized block design extended to growth and response curves. *Commun Statist. A*. 1979; 8: 191–205.

Zerbe GO; Murphy JR. On multiple comparisons in the randomization analysis of growth and response curves. *Biometrics*. 1986; 42: 795–804.

Zerbe GO; Walker SH. A randomization test for comparison of groups of growth curves with different polynomial design matricies. *Biometrics*. 1977; 33: 653–657.

Zimmerman H. Exact calculations of permutation distributions for r dependent samples. *Biometrical J*. 1985; 27: 349–352.

Zimmerman H. Exact calculations of permutation distributions for r independent samples. *Biometrical J*. 1985; 27: 431–443.

Zumbo BD. Randomization test for coupled data. *Perception & Psychophysics*. 1996; 58: 471–478.

# Index